AF333790

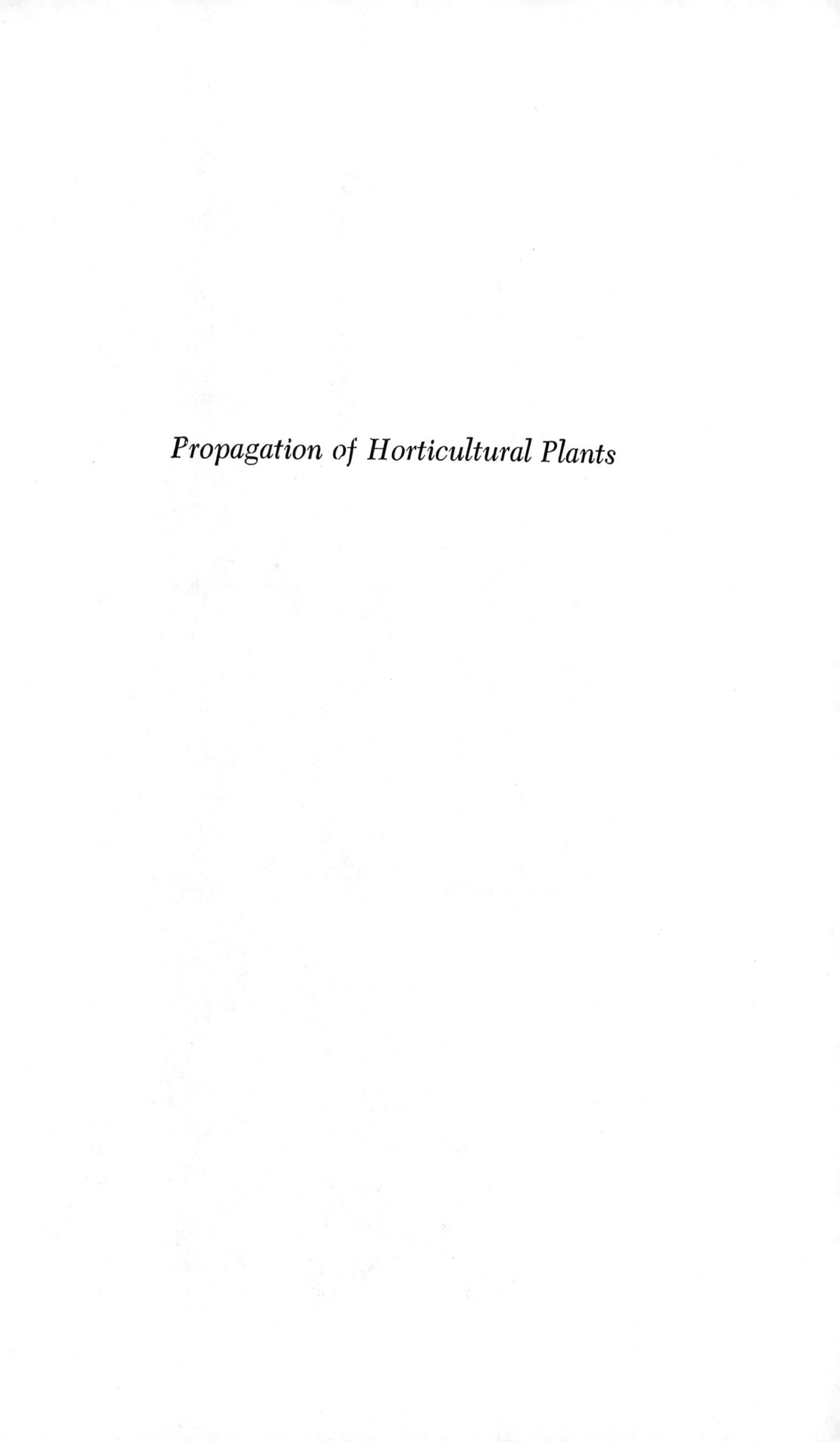

Propagation of Horticultural Plants

Orchard trees like these represent a permanent agriculture.

PROPAGATION OF HORTICULTURAL PLANTS

Guy W. Adriance

Head of Department of Horticulture, Agricultural
and Mechanical College of Texas

Fred R. Brison

Professor of Horticulture, Agricultural
and Mechanical College of Texas

SECOND EDITION

ROBERT E. KRIEGER PUBLISHING COMPANY
HUNTINGTON, NEW YORK
1979

Original Edition 1955
Reprint Edition 1979

Printed and Published by
ROBERT E. KRIEGER PUBLISHING COMPANY, INC.
645 NEW YORK AVENUE
HUNTINGTON, NEW YORK 11743

Printed in the United States of America

Library of Congress Cataloging in Publication Data

Adriance, Guy Webb, 1895-
    Propagation of horticultural plants.

    Reprint of the 2d edition published by McGraw-Hill, New York in series: McGraw-Hill publications in the agricultural sciences.
    Bibliography: p.
    Includes indes.
    1. Plant propagation.   I. Brison, Fred Robert, 1899- joint author.   II. Title.
[SB119.A3 1979]   631.5'3   79-9753
ISBN 0-88275-965-5

Preface

The propagation of plants has long been recognized as a fundamental practice in the fields of plant science. Various special treatments are followed in the production of horticultural plants, and for this reason their propagation presents certain peculiar problems and difficulties.

The original edition of the book was prepared primarily as a text for basic courses in horticulture and related fields; and also as a guide in practical work, for both commercial growers and amateur horticulturists. In the revision, considerable new material has been added, to include information that has been made available within recent years.

The essential features of plant structure and reproduction have been introduced in their relation to seed production, root formation, wound healing, and other practical phases of plant propagation. The methods of asexual propagation, including bulbs, layerage, cuttage, budding, and grafting are considered in comprehensive form. Practices followed in the propagation of certain important species are presented in detail. A discussion of pruning and transplanting is also included because of the relationship of these practices to the growth and longevity of plants (after they are planted in permanent locations). Basic factors of plant growth and response are considered with respect to their relation to commercial practices.

The material included in this work is considered to be of fundamental value to students in agriculture and especially useful to students in succeeding courses in vegetable crops, fruit growing, forestry, floriculture, and ornamentals. As a guide in practical work, the book assembles in readily accessible form recent practices that have been introduced by research workers and commercial propagators. Modifications of standard practices are reviewed and evaluated; accepted methods for the commercial propagation of specific plants are given in detail. The relative value of different rootstocks, as determined by results in many parts of the country, is given considerable emphasis.

Review of the literature has been of a general nature, and no specific references have been given. Selected references to more recent publications on the various subjects have been included, with the idea that they would in turn supply references to previous publications in the same field.

In a critical review of the various chapters, in both the original and the revised editions, invaluable service has been rendered by various members of the staff of the Department of Horticulture, who have assisted also in the preparation of photographs for illustrations. Their cooperation is acknowledged with gratitude. Photographs for illustrations have also been supplied from other surces, and credit for each has been given in the text.

GUY W. ADRIANCE
FRED R. BRISON

Contents

Preface vii

Chapter 1. History and Development of Horticulture . . . 1

2. The Structure of Plants 12

3. Inflorescences, Flowers, Fruits, and Seed . . . 26

4. Special Plant-growing Equipment 46

5. Methods of Propagation 62

6. Germination of Seeds 71

7. Methods of Seedage 92

8. Layerage 100

9. Cuttage 110

10. Bulbs and Other Modified Structures 132

11. Graftage 148

12. Methods of Grafting 170

13. Methods of Budding 188

14. Propagation of Important Horticultural Plants . . 199

15. Transplanting 260

16. Pruning 272

Index 293

CHAPTER 1

History and Development of Horticulture

The plant is the basic source of all food and consequently the determining factor in life. The fundamental process of photosynthesis, by which the plant is able to combine water and carbon dioxide to form sugar, permits the synthesis of the more complex compounds by the plant and the use of these compounds by man and the lower animals to sustain their life processes. From the earliest times, when herdsmen sought and protected grass for their herds, up to the present period of diverse plant form, man has always had a fundamental interest in the production and care of plants.

As soon as some adventurous soul found a fruit, a berry, or a plant to be edible, it immediately became an object of solicitous attention. The best types were selected and propagated by any means available, usually seeds, with the result that many new forms were constantly coming into being from which better selections could be made. Over a long period of years, this process of selection and propagation of new and better types has resulted in an array of plant materials that is lavish beyond the imaginations of earlier generations.

As man began to grow these plants in gardens, he became aware of the beauty that develops in any systematic and well-cared-for planting. He began to seek and select flowering plants, ornamental shrubs, and trees of all kinds and to blend them into pleasing landscapes. This later development of ornamental horticulture has expanded with ever-increasing enthusiasm. With all the magnificent gardens of both past and present, there is little reason to doubt that even greater achievements will be made in this field in the future.

1

History. Many passages are found in the Bible concerning horticultural plants such as the fig, the vine, the olive, the rose, the lily, and the plant that grew from the mustard seed. In the ancient days, all formal groupings of plants were called gardens. From *hortus,* the Latin word for garden, has come the term *horticulture,* which includes everything from the small planting of flowers or shrubs around the humble cottage, to famous and elaborate formal gardens; from the single plum, fig, or pecan

Fig. 1. Cherry orchard.

tree growing in the back yard, to the endless miles of apple and citrus trees wherever they are grown; from the small vegetable garden around the farm home or on the city lot, to the broad acres of onions or tomatoes grown throughout the land.

The Romans constructed magnificent aqueducts to provide water for their gardens, and Italy continues to be a paradise for those who love horticulture. France and England were also cradles for the development of horticulture in all its manifold expressions, and the early colonists brought this knowledge and interest to this country. The horticultural societies of Pennsylvania, New York, Massachusetts, Georgia, and many other states exercised a powerful influence in the early years of this country in stimulating the planting of fruit trees. Some of the early orchards would have been a credit to modern fruitgrowers, with large collections of excellent varieties.

Horticultural Industries. In modern times, horticulture has been separated into several fields of specialized interest:

Pomology. Fruitgrowing is one of the most ancient of horticultural industries, as indicated in records of many of the early

Fig. 2. Grape planting.

civilizations of the worlds. In this country, fruitgrowing has reached a very intensive stage of development, from the standpoint of technology. The commercial fruit industry of the United States, including citrus fruits, apples, peaches, grapes, pears, strawberries, cherries, plums, and many others, extends over an area of 4½ million acres with an annual average production in recent years of 16 million tons and a cash value to the growers in excess of $900 million. Tree-nut production averages annually 200,000 tons with a cash value to the growers of $70 million.

Olericulture. Vegetable production is also an ancient horti cultural enterprise. Aside from home and market gardening, vegetable production in the United States has been expanded over tremendous areas with the growing of such crops as lettuce, tomatoes, melons, onions, carrots, cabbage, and many other truck crops, for shipment by railway and truck refrigeration to distant markets. The commercial vegetable industry covers approximately 3¼ million acres with an annual average production

Fig. 3. Tropical garden (vegetables and bananas).

of 14 million tons and a cash value to the growers of $800 million.

Floriculture and Ornamental Horticulture. The commercial production of flowers and the production of trees and shrubs for landscape planting are much more recent developments, but both have reached tremendous proportions in a short period of time. There are at present approximately 5,000 major growers of floral crops and 11,000 others operating on a lesser scale, employing 150,000 people and utilizing 200 million square feet of greenhouse space. The wholesale floriculture crop amounts to $330 million and the retail business to $650 million. The nurseries of the United States have a capital investment of $110 million and an average annual crop value of $71 million.

Throughout the world, there are many horticultural industries of outstanding interest and significance, such as the bulb industry of Holland, coffee in Brazil and Central America, bananas in Central and South America, and cacao in several of the tropical areas, principally in Asia and Africa. In the case of citrus fruits,

Fig. 4. Pineapple plantation.

world production, representing many countries, reached the staggering total of 400 million boxes in 1952–1953.

In connection with these major phases of production based on type of plant material involved, there are also several important kinds of work that involve various plants in each group.

Propagation of plants is accomplished by many different methods, of which seedage is the most common; other methods include the use of various plant parts such as bulbs, rhizomes or tubers, layers or cuttings, and finally budding and grafting. These practices are fundamental in the many types of greenhouse and nursery industries; and even in the case of seedage many new techniques are constantly being developed to aid growers.

Processing of horticultural products by dehydration, pickling, canning, and quick freezing represents another industry that has expanded rapidly in recent years. For example, the canned pack of three vegetables and three fruits in a recent average season was 127 million cases. The pack of frozen vegetables was 895 million pounds; of only three fruits, 300 million pounds; and of frozen citrus concentrates, 50 million gallons.

Breeding of horticultural crops and testing of varieties are other types of work which continue to be of increasing import-

Fig. 5. Well-trained trees in a commercial nursery.

ance from year to year. New materials are being continually introduced from other countries and new methods of inducing variations have resulted in a much wider range of characters for selection. A few years ago, most fruit varieties were the result of chance seedlings, but each year an increasing number of new varieties with desirable characters are being made available as a direct result of this breeding program.

Classes of Plants. There are many different systems of classifying plants, from the standpoint of botanical relationships and stage of development. As viewed by the horticulturist, however, plants may be classified according to the relative length of time that they require to complete a cycle of growth.

Life Cycle of Plants. The life cycle of any plant is the entire process of growth that is involved from the germination of the seed to the production of a crop of seed by the new plants. The actual duration of this period is very variable, since some plants mature and produce seed much more quickly than others. Portulaca and some of the grasses finish the life cycle in a few weeks, while the Northern Spy apple does not bear fruit and make seed until ten to fifteen years old.

The behavior of certain plants, from the standpoint of the life cycle, is quite different from that of others. One group of plants, known as *monocarpic,* produces only one crop of seed and then dies. In many of these plants the length of life may be prolonged by preventing them from flowering and producing seed. Other plants, which may also be said to have completed their life cycle when they produce seed, do not die but continue to live and produce seeds for many years.

Annuals. These are plants that mature seed during the same season in which the seedlings are started. Many annuals, such as zinnia, produce their seed early enough to make a second stand of plants in the latter part of the growing season. Corn, bean, and watermelon are examples of annual plants. Some plants fall into this group of annuals as the result of certain conditions of environment. Cotton lives many years in its native environment but is an annual under the climatic conditions of the United States. The papaya in Texas and Florida also becomes an annual in some cases. Plants from other groups may come in the annual group under other conditions. With the long growing season of the South, many biennials complete their life cycle in one season. Other biennials, of which Canterburybell is a notable example, have been changed over to annuals by the efforts of plant breeders.

Biennials. Plants that make one season's growth, survive the winter in the dormant condition, and produce seedstalks and seed the second year are known as biennials. The entire plant may go through the winter, or it may die back to the root and come up again. The plants normally produce seed and die before the end of the second season of growth. In many cases, biennial plants run to seed the first year and do not last over into the second season. Mustard, cabbage, and hollyhock are examples of biennials.

Perennials. These plants persist from year to year, often not producing seed for many years, but usually not dying immediately after seed production. Fruit orchards, especially nut orchards, produce seeds year after year, over a long period of time. There are three general classes of this group:

Herbaceous perennials have tops that die down every season, while the roots persist. These plants are typified by mallow, asparagus, and Queen's crown vine.

Woody perennials have a perennial root also, but the tops last two seasons instead of one. Dewberries and blackberries are of this type; the canes that are produced one season will fruit and die the next.

Trees and shrubs are completely woody plants, which persist from year to year, adding to that portion of the plant body previously produced. Each of the annual rings in the wood of trees cut transversely normally represents one year's contribution to the body of the plant.

Plant Dissemination. The origin of plants was probably restricted to a few large zones or regions, but plants themselves are good travelers, and in the course of their travels, their origins and even their original forms have become considerably obscured. Modifications in flower structure and function have resulted in cross-fertilization and the consequent production of unlike individuals; and this in turn has enabled species and varieties to spread into new areas where the original forms would not have been able to survive.

Many plants have structural modifications of fruits, seeds, or other parts that facilitate or encourage dispersal. The extent to which plants possess such modification determines their mobility. Those which have large heavy fruits or seeds are relatively immobile; those which produce light fruits or seeds, equipped perhaps with wings or similar appendages, have high mobility. Dissemination, or migration, depends upon the mobility of plants and also upon various agencies or factors that tend to move plants from one place to another. Those species which, under natural conditions, have the best facilities for distribution are likely to be found over the greatest range of territory. Oceans, deserts, and mountain ranges act as natural barriers to prevent plant migration by natural means. As a general rule the distribution of a species is also dependent upon its age. The pecan is undoubtedly of comparatively recent origin, as shown by its limited distribution. It is not to be supposed that the small area of this country in which it grows wild is the only region to which it is adapted.

Plant Introduction. Man either purposely or by chance is often responsible for the distribution of plants. The American colonists very early introduced into America plants that had been grown in the countries from which they came. The U.S. Department of Agriculture maintains plant explorers in several foreign countries, looking both for new fruits and for more desirable forms of those which are now available. Valuable horticultural plants have been introduced by these explorers, many of them coming from China and South and Central America and many others from Europe and Western Asia. Material secured is tested in government introduction gardens before being distributed widely. Some of the most useful fruit and vegetable plants are native to America, the principal ones being the tomato, potato, corn, bean, strawberry, grape, blueberry, cranberry, plum, crabapple, and pecan. Several others, however, such as the cultivated apple, pear, peach, cherry, citrus fruits, date, fig, and most species of vegetables have been introduced from foreign countries. Introductions are frequently valuable in themselves, but more often their chief value is for breeding work.

Wind. Wind is probably the most effective natural agency in the dissemination of plants. The seeds and fruits of many plants have special structures and modifications that enable the wind to carry them. The fruits of maple and elm have wings, and those of the lettuce have hairy disseminules that enable the seeds to be borne by the wind. In other plants the seed is modified so as to be blown about easily. Examples of these are the willow, milkweed, cotton, and cattail. Tumbleweeds, which roll across the prairies as whole plants, illustrate a special case of wind dissemination. The Russian thistle is a plant of this kind. The plant breaks off at the ground after frost and may be blown a great distance, frequently 50 to 75 miles, during the winter months. The small seeds are so enclosed that they do not all fall out, but instead are scattered along the path of the plant as it is swept along by the wind.

Water. Ocean currents, running streams, and lakes are means whereby plants are scattered. Seeds and fruits have various modifications that enable them to float. The coconut, with its thick, fibrous husk and hard shell, is an illustration. Many nuts, such as the oak, pecan, and walnut, are readily distributed by water. Seeds of grasses and noxious weeds are often carried downstream by overflows of rivers or creeks.

Animals. Seed may be distributed by animals in three principal ways: by carrying them in fur, hair, or feathers; by eating the seeds and passing them through their digesitve systems; or by storing or burying them. Cockleburs, clover, and needlegrass readily become attached to the hair or hide of animals. The fruit of Martynia, commonly known as devilsclaws, splits in a way that forms two incurving hooks, which readily attach themselves to animals. The seeds of many edible fruits, such as peach, plum, and berries, are carried in the digestive system of certain animals. Nuts are frequently transported by squirrels, blue jays, and crows and dropped at a considerable distance from their point of origin.

In a few cases, plants are disseminated by parts other than seed. This is true of Bermudagrass, which is easily carried from one place to another attached to the feet of the animals.

Propulsion. The fruits of some plants are characterized by an explosive action when ripe, and the seeds are thereby scattered. The distance to which they are propelled may not be great, and the rate of distribution is slow, but when continued indefinitely the effect becomes of importance. Propulsion of seeds is due to the unequal drying of different layers of the ovary wall, resulting in strains that are suddenly overcome when the fruit explodes. The fruits of bull nettle, violet, "sandbox," castorbean, and cowpea illustrate this phenomenon.

QUESTIONS

1. What is the meaning of the word horticulture?
2. Define pomology, olericulture, floriculture.
3. What is the life cycle of a plant?
4. What is the difference between a biennial and a perennial?
5. What is meant by a herbaceous plant?
6. Name some important horticultural plants that are native to America.
7. From what countries have many of the horticultural plants been introduced?
8. What is the most important plant part from the standpoint of

dissemination? What are the principal agents in the dissemination of plants?

SUGGESTED REFERENCES

Boswell, Victor R., and Else Bostelmann: "Our Vegetable Travelers," *National Geographic Magazine*, **96**:145–217, 1949.

Edmond, J. B., A. M. Musser, and F. S. Andrews: "Fundamentals of Horticulture," The Blakiston Company, 1951.

Klages, K. H. W.: "Ecological Crop Geography," The Macmillan Company, 1942.

Magness, J. R., and Else Bostelmann: "How Fruit Came to America," *National Geographic Magazine*, **100**:327–377, 1951.

Shoemaker, J. S.: "General Horticulture," J. B. Lippincott Company, 1952.

Talbert, J. T.: "General Horticulture," Lea & Febiger, 1946.

CHAPTER 2

The Structure of Plants

The propagation, culture, and management of horticultural plants are based to a considerable extent upon a knowledge of the structure of the plant and the function of its different parts. The larger structural units of the plant, which are known as organs, are the roots, stems, leaves, flowers, and fruits. Each is composed of several different kinds of tissues, such as xylem, phloem, and cambium, and these tissues, in turn, are composed of cells. Some cells have thick walls, others have thin walls; cells differ also in size, shape, and cell contents. The three principal types of cells in plants are parenchyma, schlerenchyma, and collenchyma.

Roots. The roots are essential organs of most plants. The chief functions of roots are to absorb moisture and nutrients for the plant and to provide anchor for it. The following kinds of roots may be conveniently considered from the standpoint of origin, structure, and functions:

Primary Roots. The radicle of a germinating seed produces the first root of the new plant. This is the primary root. Its general direction of growth is downward. The continued growth of this primary root produces the so-called "taproot." In some plants, as walnut and hickory, the growth of the taproot predominates for several years, and they are commonly regarded as taprooted plants.

Secondary and Lateral Roots. Branch roots that arise from the taproot are known as *secondary* roots. The general direction of growth of secondary roots is horizontal. Roots that develop laterally on any previously formed root are known as *lateral* roots. In reality, these may develop from the taproot, from other

12

lateral roots, or in some cases from stems. There is considerable variability in the extent of branching shown by roots of different species of plants. The tomato is an example of a plant in which free branching occurs; root branching in the onion, on the contrary, occurs less freely; and the hyacinth produces roots that are normally unbranched.

The peach and apple are examples of plants in which there is limited development of the taproots but extensive development of the lateral roots. Plants in which this occurs are known as lateral-rooted plants. The spread and depth of the root system and the extent of branching are also influenced by methods of propagation, moisture, and soil. Free branching normally occurs in fertile soil, and limited branching in poor soil. Hence plants in sandy soil of low fertility tend to produce long roots with relatively few branch roots.

Regular Roots. As the radicle of the germinating seed begins to grow, it consists initially of primary cells which form primary tissues. As it and the branch roots that develop from it continue to grow, the region extending a short distance from the tips is char-- acterized by primary growth. In cross section it has an epidermis on the outside, and successively inward, cortex, endodermis, pericycle, phloem, cambium, and xylem elements. Secondary growth of the roots of some plants is initiated as the roots become older, and a cross section will consist only of periderm, phloem, cambium, and xylem. Thus, during growth a given root will consist of a basal portion which is characterized by secondary growth and a terminal portion which is characterized by primary growth. Branch roots arise from the pericycle tissue, at a position shortly back of the growing tip. The youngest roots are always nearest the tip of the root, and the older ones are toward the base. Since they thus develop in regular succession, they are known as *regular* or *acropetal* roots.

Adventive Roots. Those that arise from other tissues and organs of the plant than the pericycle of young roots are called *adventitious* or *adventive*. Briefly, adventive roots may form in varying degrees of readiness, from roots, stems, leaves, and modified parts of the principal kinds of horticultural plants. They never, or rarely, form from the *roots* of *monocotyledonous* plants, but form readily from the *stems* of certain ones; they form readily from the *roots* and *stems* of some *dicotyledonous* plants, but less readily from others; they form quite readily from the leaves of certain *monocotyledonous* and *dicotyledonous* plants, but not from *gymnosperms*. The tissues from which adventive roots originate are principally the cambium layer of roots, the cortex, pericycle, phloem, cambium, and callus of stems, and the parenchyma tissue near the cambium of vascular bundles of leaves.

Root Hairs. Simple, hairlike outgrowths of the outer walls of cells of the epidermis of the root are produced by many plants. These are known as *root hairs.* They grow out into space

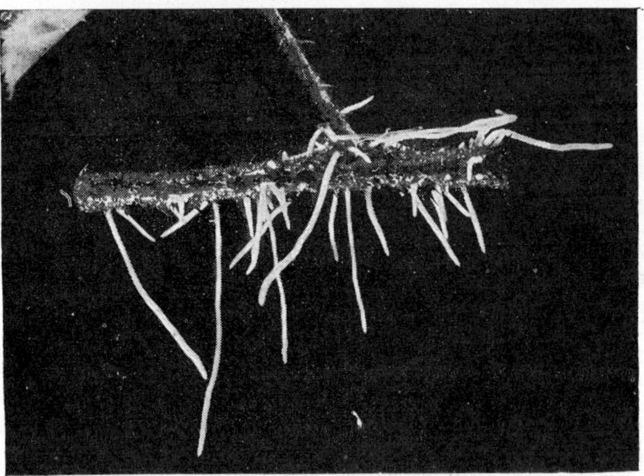

Fig. 6. Adventive roots—blackberry.

between soil particles and absorb moisture and nutrients for the plant. Most of the higher vascular plants, such as peach, apple, and blackberry, have root hairs. Some, such as the orange, grapefruit, cranberry, and pecan, do not have normal root hairs, at least under certain soil conditions. With these the absorption of moisture is performed by various small lateral roots. They function for a short time as absorbing organs; then they either die or begin secondary growth and become a part of the permanent root system. Since the epidermis is present only on those portions of the root which consist of primary tissue, it is obvious that the root hairs occur only on the terminal portion of young growing roots. Root hairs normally function only during a relatively short period. As they wither and disappear, others develop near the terminal growing point of the root.

Stem. In the germination of a seed the plumule produces the first stem of the new plant. The continued growth for one or more seasons of this first stem and, in most species, the branching of it produce the trunk and framework of the new plant. Stems, in turn, produce the buds, leaves, flowers, and fruit. They

Fig. 7. A well-developed root system showing tap and laterals. (*Courtesy of O. S. Gray Nursery, Arlington, Texas.*)

also serve as conducting systems for water and nutrients between the roots, leaves, and fruit.

The principal groups of horticultural plants are angiosperms (including monocotyledons and dicotyledons) and gymnosperms.

Monocotyledonous Plants. In general structure the stems of these plants consist of a terminal growing point, nodes, buds, and internodes. In cross section the stem consists of epidermis, perhaps a cylinder of thick-walled schlerenchyma beneath the epidermis, and isolated vascular bundles distributed in a mass of fundamental tissue similar to the pith of dicotyledons.

The internodes of monocotyledonous plants provide most of the length growth of the stem. The lower part of each internode

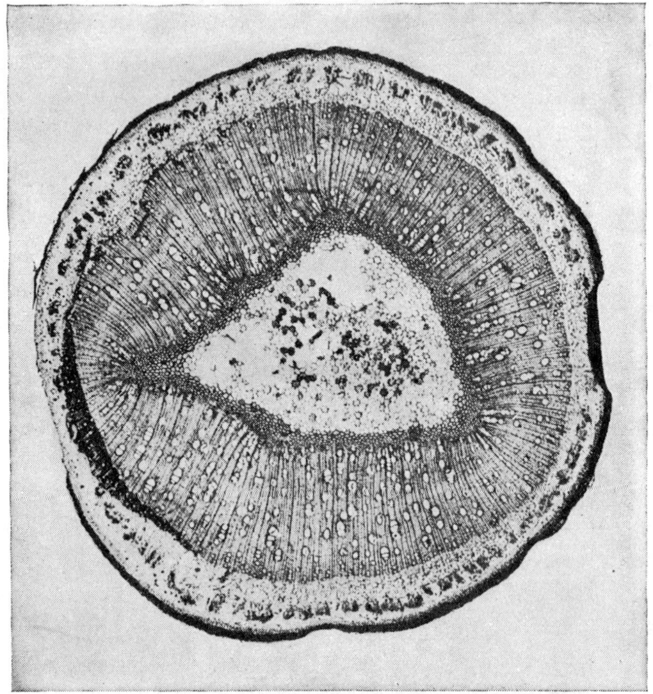

Fig. 8. Cross section of woody stem, showing pith; xylem with medullary rays and water-conducting vessels; and bark which includes cortex and phloem.

is a region of elongation. This in part accounts for the exceedingly rapid length growth of stems of certain monocotyledonous plants, such as the bamboo, Johnsongrass, and asparagus.

The elongation of the stem may continue for many years. The date palm, for example, lives for an indefinite period and may ultimately grow to be 50 feet high; corn, on the contrary, survives only one season and rarely grows taller than 6 to 8 feet. In some species the terminal growing point produces the inflorescence of the plant. This is true of the onion, and also of corn, which produces the tassel terminally on the main stalk and the pistillate inflorescence on a lateral branch.

The nodes of monocotyledons sometimes give rise to axillary buds. Such buds are rare on the date palm, with the result that the plants normally produce single unbranched stems with only occasional offshoots. Corn, on the contrary, produces buds freely at the nodes, some of which grow into branch stems while others produce the ears of corn.

Dicotyledonous and Gymnosperm Plants. Young stems of these plants have a pith, xylem, cambium, phloem, pericycle,

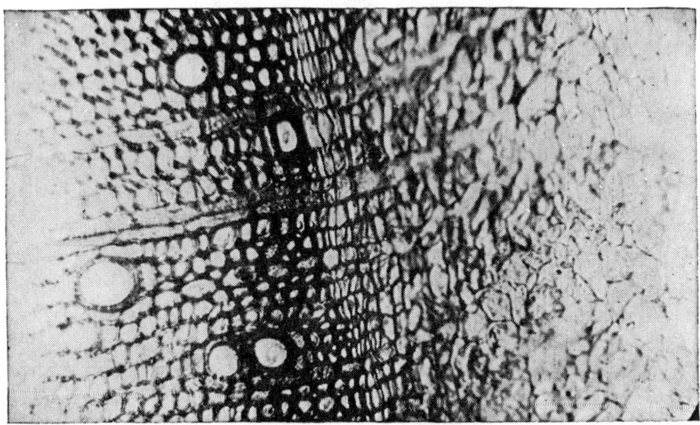

FIG. 9. Section of woody stem, greatly enlarged to show cambial cells.

endodermis, cortex, and epidermis in successive order outward from the center. In old stems the epidermis and cortex have disappeared, and exposed phloem cells form a periderm, or bark. Stems of dicotyledons make terminal growth by elongation of cells near the tip of a growing branch. The terminal growing point in its process of growth may (1) produce a terminal bud, from which growth will be resumed the following season, (2) produce a terminal inflorescence as in the grape, apple, pecan, walnut, cabbage, and carrot, or (3) it may abort, in which case future growth of the stem will be from an axillary bud below.

Young stems in the process of growth and development differentiate into nodes and internodes. Leaves and buds are normally formed at nodes. The area between nodes is known as the *internode.* The internode in stems may be very short, as in cabbage, or relatively long, as in the grape, depending on the species. Rapidity of growth caused by growing conditions also influences the length of the internodes of many species.

The cambium layer of stems of these plants is of peculiar interest and concern because of its relationship to several im-

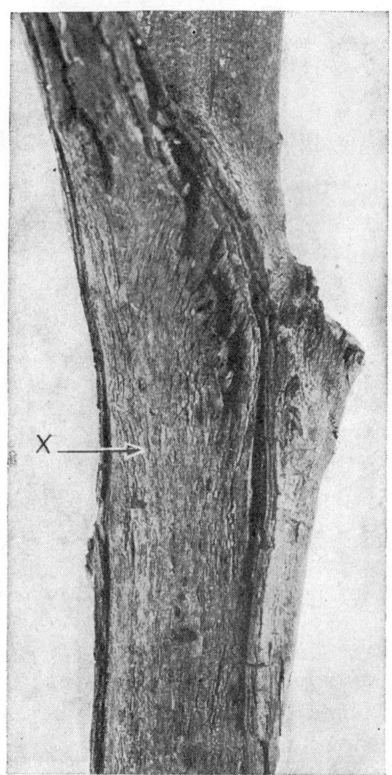

FIG. 10. The tissue shown in the area marked by X developed by regeneration from the injured cambium when the original bark was peeled off.

portant horticultural practices and treatments. Briefly, the cambium layer serves the plant in these ways: (1) It is the meristematic tissue responsible for increase in size of the stem after it begins secondary growth. Cambium cells occur in a continuous ring between the xylem and phloem. During each season of growth they enlarge and divide and differentiate into new xylem cells toward the inside and new phloem cells toward the outside. If the cambium fails to function because of mechanical injury or physiological causes, no xylem and phloem are formed. Since these tissues are essential to the normal growth of a plant, death results if the activity of the cambium layer is restricted for a prolonged period. (2) The healing of wounds is made possible by the cambium layer and is accomplished by

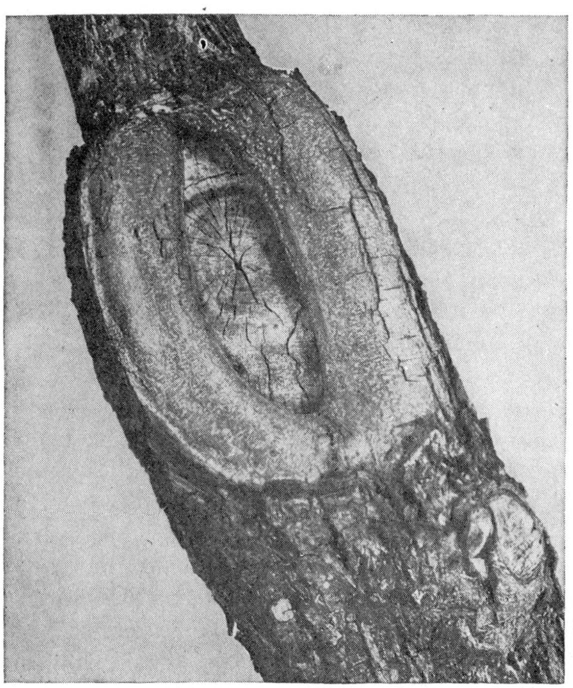

Fig. 11. Healing of a wound by overwalling. No wound compound was applied and the wood checked as a result of exposure.

two processes: *Regeneration* may take place where bark is removed and living cambium cells are exposed on the surface of the wood. Under the favorable conditions of high humidity and warm temperature such cells may become active and reconstruct new tissues on the surface of the wound. New growth from these cells is outward in a radial direction. *Overwalling* takes place as a result of the growth in a lateral direction of cambium cells around the margins of a wound, causing new tissues to advance from various sides to cover the wound. (3)

The cambium produces *callus* tissue which is essential to the success of budding and grafting. Callus also forms on the cut ends of cuttings of some plants. This may provide protection against decay-producing organisms. In rare cases, roots arise directly from the callus tissue, though in most cases they arise directly from the cambium and callus is not essential to rooting. (4) Finally, when adventitious roots develop on stems or on roots, they usually arise in the cambium or from recently differentiated cells near the cambium.

Buds. A bud is a growing point, surrounded by small, partially developed leaves. It is in reality a rudimentary stem in a state of dormancy or limited growth, protected by an envelope of bud scales. It may consist of a mass of meristematic cells or of several nodes and very short internodes. Close examination of a well-developed bud reveals leaves and buds in the same order as on a growing stem of the same plant. Several classifications of buds are recognized, based principally on the mode or time of origin, position on stem, position on node, time at which they begin growth, and function.

Position on Stem. In the growth of stems, buds are formed at different positions. The principal kinds of buds, with regard to their location on the stem, are *terminal, axillary,* and *lateral.* Terminal buds are those that develop from the terminal growing point at the end of a stem when growth ceases. In some kinds of plants, they are formed regularly; in others, the growing point tends to abort, leaving no bud. The shoots that develop terminal flowers or inflorescences do not produce terminal buds. In the event that a terminal bud does form, it is usually the one to begin growth first the following spring. A terminal bud is regarded as being dormant or largely so; whereas a *terminal growing point* is regarded as being in a state of active growth and elongation. An axillary bud is one that occurs in the axil of a leaf—the angle between the leaf and the stem. They are designated as axillary buds even after the leaf has shed. These buds are also properly called *lateral* buds, because they occur on the side of the stem. Lateral buds, however, may occasionally occur at nodes where no leaf occurs or where the leaf was rudimentary. Examples of the latter class are frequently observed on parts of pecan shoots that are formed near the end of the growing season.

The peach, tung tree, and less frequently the rose produce shoots with certain nodes and leaf axils at which no buds occur. These are commonly called *blind* buds or *blind* nodes.

Position at Node. In most species only a single bud develops in the axil of each leaf. In some, however, two, three, or even

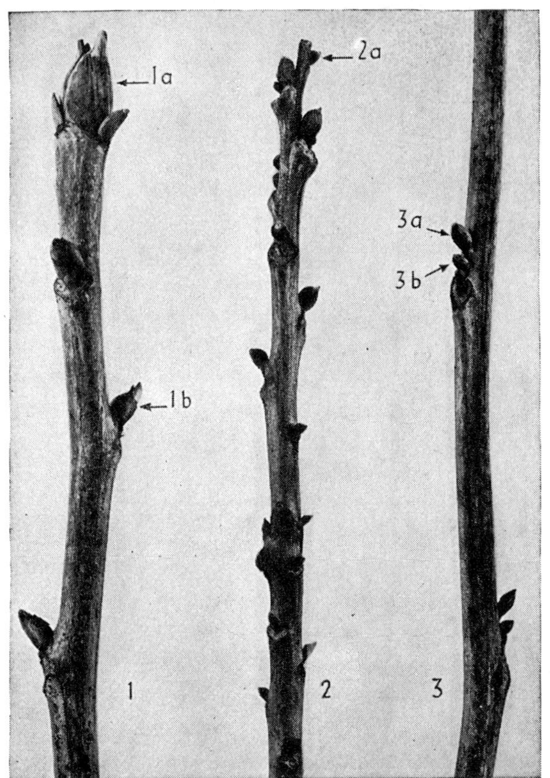

Fig. 12. Buds on one year stems. 1, Hickory, showing (1*a*) terminal bud and (1*b*) lateral bud; 2, Pecan, showing (2*a*) lateral bud in terminal position; 3, Pecan, showing (3*a*) primary bud and (3*b*) secondary bud.

more may develop. The bud nearest the terminal of the shoot is usually the largest of the group and is named the *primary* bud. The next bud, then, becomes the secondary bud, and so on. Commonly, however, all except the primary bud are referred to collectively as the *secondary,* or *reserve,* buds.

The primary bud is the one of the group most likely to grow

when the tree starts growth in the spring. The reserve buds
oftentimes begin normal growth with the primary buds, but
they are especially likely to grow under conditions of excessive
soil moisture or if the growth from the primary bud is injured
by cold weather, insects, or other causes.

Fig. 13. Blind nodes of peach, between terminal and lateral buds which had
begun growth.

Vegetative and Flower Buds. The growth in height of a plant
and the production of branches is due to the growth of *vegetative*
buds. These are also called *leaf* buds.

Flower buds contain the rudimentary blossoms with various
parts of the flower enclosed. Since flowers normally produce
fruit as they continue to grow and develop, these buds are also
known as *fruit* buds. Flower buds develop from, or in close

association with, vegetative buds; hence, they occur on plants in the same general position as vegetative buds. In some species, they can be readily distinguished from vegetative buds; in others, the two are quite similar in appearance. The formation of flower buds takes place in some species during the season previous to the one in which the flowers appear. In other species the flower buds do not form until a time shortly before the buds begin to grow. In the peach, for example, the flower buds that bloom in the spring are formed during the previous summer and fall. Flower buds of citrus develop in late winter or early spring preceding the blooming period. An accumulation of *stored food* in a plant is regarded as favorable for fruit-bud formation, and this accounts for the considerable variation in the time within a species when flower buds are formed.

Some species of plants produce *mixed* buds. These contain both flowers and vegetative parts within the same bud; consequently, when they begin growth they produce both vegetative growth and flowers. The apple, pear, and blackberry are plants that produce mixed buds.

Dormant and Latent Buds. The buds of most fruits develop and mature during a given season and remain *dormant* over winter. Such buds begin growth the following spring and either develop into shoots or fruits or fall off, or they may remain dormant for a period of one to several years, in which case they are called *latent* buds. They may even become covered over by layers of bark; however, these latent buds usually make sufficient annual growth outward to prevent them from being overwalled. When trees are cut back heavily, any of the "water sprouts" that develop arise from latent buds.

Adventitious Buds. Normally shoots arise from well-formed buds, but occasionally they develop from other tissues, which form *adventitious* buds, and shoots that grow from these are called *adventive.* Those that arise from roots are known as "suckers"; for example, buds of the pear, blackberry, and persimmon plants. The point of their origin is in the cambium of roots. Adventive shoots may also arise from stems. These originate principally in the cambium layer. Those that occur on the body or framework of trees and which make rank, vigorous growth are called "water sprouts." Most such shoots, however, arise from latent buds. Adventive shoots are also produced readily by leaves of some plants and less readily by others. The shoots originate from parenchyma tissue close to the vascular bundles in leaves of dicotyledonous plants; and frequently from *callus*, formed at the cut or injured portion of the leaves of monocotyledonous plants.

Leaves. Mineral nutrients and water from the soil are combined with carbon dioxide in the leaves, under the influence of sunlight, to form plant foods essential to growth. Leaves are lateral appendages formed by the stem in elongation. Plants that shed their leaves at certain seasons, and hence have a period during each yearly cycle when they are bare and another when they are in full foliage, are known as *deciduous* plants. Examples are peach and apple. Those that retain their leaves for long periods, and do not shed all of them at one time but shed them so gradually that the trees have leaves on them at every season of the year, are known as *evergreen* plants. *Broad-leaved* evergreens are represented by such plants as citrus and cherry-laurel, which retain their leaves for more than one year, and by others such as the live oak and yaupon, which retain them only until new leaves are formed the following spring. The pines, arborvitaes, spruces, and junipers are examples of the *coniferous* evergreens.

QUESTIONS

1. What are the larger structural units of a plant?
2. What are the chief functions of the roots of a plant?
3. Name functions of the part of a root near the tip that consists of primary tissue.
4. Name functions of the part of the root that consists of secondary tissue.
5. How do root hairs arise? What is their function? How long do they live?
6. In what ways do branch roots arise?
7. What is the function of the stem of a plant?
8. Where is the cambium located? Of what value to the plant is it?
9. What is callus? Regeneration? Overwalling?
10. Define a bud.
11. Classify buds according to position on stem, position at node, and function.
12. What is the difference between a sucker and a water sprout?
13. Distinguish between deciduous and evergreen plants. What are common examples of each kind?

SUGGESTED REFERENCES

Eames, A. J., and L. H. MacDaniels: "An Introduction to Plant Anatomy," McGraw-Hill Book Company, Inc., 1925.

Hayward, H. E.: "The Structure of Economic Plants," The Macmillan Company, 1948.

Hill, J. B., L. O. Overholts, and H. W. Popp: "Botany: A Textbook for Colleges," McGraw-Hill Book Company, Inc., 1936.

MacDaniels, L. H.: Anatomical Basis of So-called Adventitious Buds in the Apple, *Cornell Univ. Agr. Sta. Mem.* 325, 1953.

Strasburger's "Textbook of Botany" (rewritten by Fitting, Jost, Schench, and Karsten), Macmillan & Co., Ltd., London, 1920.

Weaver, J. E., and W. E. Bruner: "Root Development of Vegetable Crops," McGraw-Hill Book Company, Inc., 1927.

CHAPTER 3

Inflorescences, Flowers, Fruits, and Seed

Some horticultural plants are grown primarily for their vegetative parts. Others are grown for the flowers, fruits, or seed which they produce.

Inflorescences. Although the flowers of many plants are borne singly on stalks or stems known as pedicels, in numerous other cases the flowers are borne in clusters known as *inflorescences*. The principal parts of the *inflorescence* are the peduncle, pedicels, and individual flowers.

The *peduncle* is the main stem or central axis. From it arise the *pedicels* or in many cases the flowers directly without pedicels. A flower, then, may be borne on a pedicel or it may be attached to the main axis or peduncle without any stalk or pedical, in which case it is known as *sessile*. The area of attachment of the flower to the pedicel, or in the absence of a pedicel, to the peduncle is known as the *receptacle*.

There are several distinct types of inflorescences, such as spike, raceme, corymb, head, fascicle, and glomerule, based upon the positions and relationships of the different parts. A *determinate* inflorescence, for example, is one in which the inflorescence terminates in a flower, as in the apple, and an *indeterminate* inflorescence is one in which the terminal remains vegetative, with flowers borne laterally, as in the cabbage and hyacinth. In some cases the peduncle is branched, giving rise to a compound inflorescence. In some flower clusters both simple and compound types are represented. This is true in the grape and cabbage.

Flowers. The flower is the forerunner of the fruit and seed. In order to consider the processes that result in fruit and seed

26

FIG. 14. Inflorescence of pear showing flowers borne on pedicels which arise from the peduncle.

FIG. 15. Inflorescence of onion consisting of many pediceled flowers borne terminally on peduncle.

formation, it is important to give some thought to the structure of flowers. There are two essential parts of the flower: the pistil and the stamen. The pistil consists of the stigma, the style, and the ovary, which is the lower, enlarged portion. In it are borne the ovules, which, when mature, become seeds. The style forms the connection between the ovary and stigma, and through it the pollen tube passes on its way into the ovary. The stigma represents the upper portion of the pistil; its receives the pollen and affords a favorable medium for its germination. The stamen is made up of the filament, or stalk, and the anther, in which the pollen grains are produced. Enclosing these essential parts in two outer whorls are the corolla and the calyx. The corolla is made up of petals and the calyx of sepals. The calyx and corolla

Fig. 16. Two kinds of muskmelon flowers. The staminate flower at X is borne in axil of leaf A; this flower falls off shortly after it sheds pollen. The pistillate flower Z is borne in axil of leaf B; it contains stamens, and also a pistil, which develops into the fruit. Tendrils arising at the nodes where the flowers occur are shown at Y.

are accessory parts but not essential to the formation of seeds. It is considered that their function is to attract insects, some of which are helpful in pollination; they possibly afford protection to some flowers. The calyx and corolla are absent or rudimentary in some flowers; in others they have grown together to form the perianth, as in the pecan.

In some plants the flowers and fruits develop directly from a *growing point* of the plant. This happens with the tomato, muskmelon, and many herbaceous flowering plants. In other plants the flowers and fruits develop from previously formed *flower* or *fruit* buds. Two general classes of flowers are recognized.

Perfect. Those flowers that have both stamens and pistils and are, hence, generally capable of self-pollination are known as perfect or hermaphroditic flowers. Perfect flowers are borne in most of the common fruit and vegetable plants. Such fruits as the apple, plum, orange, grapefruit, and lemon have perfect flowers; the tomato, bean, lettuce, garden pea, onion, and many other vegetables likewise have perfect flowers.

FIG. 17. Two kinds of watermelon flowers. The flower at Z is staminate and will fall off after it sheds pollen. The pistillate flower shown at X will be pollinated later, and the enlarged basal portion will develop into the fruit.

Imperfect. Species that have only stamens or pistils within a flower are divided into two groups:

MONOECIOUS species are those that have the stamens and pistils borne on the same plant but in different flowers. Examples of plants of this group are the hickory, pecan, walnut, filbert, oak, and tung tree. Vegetables that are monoecious are the cucumber, squash, pumpkin, watermelon, and sweet corn.

DIOECIOUS species are those in which stamen-bearing flowers, and pistil-bearing flowers are borne on separate plants, producing so-called *staminate* and *pistillate* plants, respectively. The

date palm, fig, American persimmon, muscadine grape, asparagus, and spinach are examples of plants of this group.

Pollination. An essential step in the development of seed from
a flower is the process of pollination. When an anther is
mature, it normally splits along longitudinal grooves and its
pollen grains are discharged. This process is known as *dehiscence*. Pollination is the transfer of pollen grain from an anther
to the stigma of a pistil.

Self-pollination. Technically, self-pollination is the transfer of
pollen from an anther to a pistil of the same flower. According
to broader usage, however, *selfing* refers to the pollination of a

Fig. 18. Inflorescence of a staminate papaya plant.

pistil with pollen from a flower of the same genetic composition.
Thus, the pistil and pollen involved in selfing might be borne
by the same flower, by different flowers of the same plant, or
by flowers of different plants of the same variety.

Cross-pollination. Technically, cross-pollination is the transfer
of pollen from an anther to a pistil of any other flower. According to current usage, however, *crossing* refers to the pollination
of a pistil by pollen from a plant of a different genetic composition. Thus, the *Earliness* blackberry was developed by crossing
the Ness variety with the Louisiana dewberry variety.

The structure of perfect flowers determines to a marked degree whether pollen for pollination is supplied by the same
flower or a different flower. In the bean, for example, the pistils

and stamens are enclosed in a complex floral arrangement; pollen is easily supplied by the anthers; it is in direct contact with the stigma of the pistil, and only rarely is pollen from other flowers involved in pollination. In other cases, such as the apple and tomato, pistils may be pollinated with pollen from the same

FIG. 19. Pistillate and staminate inflorescence of walnut. Catkins borne singly from secondary buds on one-year wood; pistillate flowers in spike terminating new growth.

flowers, or readily with that brought in by bees or other insects from other flowers.

Pistils of imperfect flowers, obviously, can be pollinated only with pollen from other flowers. If the pollen comes from flowers of the same plant or from plants of the same variety, the process would be *selfing;* if it came from plants of a different variety, the process would be *crossing.*

Agents of Pollination. There are various ways by which pollen may be transferred from the anther of a flower to the stigma of the pistil.

Growth Processes. In some plants the pollen is brought into contact with the stigmatic surface of the pistil in the process of

Fig. 20. Flowers of pecan. Three-stalked catkins borne in pairs from buds on one-year wood; pistillate spike borne terminally on new growth.

the growth and development of the flower. For example, in lettuce, at the beginning of anthesis, the pistil grows through the staminal cone and pollination takes place.

Insects. Flowers of some plants have large showy petals and nectar glands within the flower that attracts insects. The pollen grains of such flowers are generally sticky, heavy, and in some cases covered with an oily film. This is true of the pollen of flowers of such common horticultural plants as the peach, plum, cherry, apple, pear, tomato, and muskmelon.

In the tomato, for example, the pollen is sticky and insects are helpful in transferring the pollen even the short distance from the anther to the pistil. Pollen will also adhere to the body of the insect for possible transfer to a blossom to be visited later. Likewise in certain monoecious plants such as the cucumber, and also in dioecious plants as the Smyrna fig, where the staminate and pistillate flowers are widely separated, insects that

Fig. 21. Honeybee visiting apple blossom. (*Courtesy of Lee Jenkins, University of Missouri.*)

visit the flowers are helpful and almost necessary in providing for the transfer of pollen.

Many different kinds of insects are instrumental in pollination. Some of the most common ones are:

HONEYBEES. Of all the insects, common honeybees are by far the most important for pollination purposes. They have a body structure especially adapted to the carrying of pollen; they become active in the early spring when most fruits bloom, and continue throughout the growing season; they survive the winters in great numbers; they are constant for only one kind of a plant at a time and visit no others during a given period; and finally, they can be moved about and made available wherever they are needed.

Fruitgrowers in sections where there are few wild bees often place hives of bees in the orchard during the period in early spring when they are needed for pollination purposes. In the greenhouse culture of American varieties of cucumbers, special provisions must be made for pollination. The pistillate flowers may be hand-pollinated, but a more successful way involves the use of bees. The hives are kept either on the outside of the greenhouse, usually with a pane of glass removed, and individual hives so placed that the bees enter the house through the opening; or the hives may be located inside the house.

FIG WASP. The Smyrna fig is dioecious and the flowers must be pollinated with pollen from Capri figs for proper development. Commercial production of these Smyrna figs depends upon the so-called "fig wasp" for pollination, and apparently no other insect is suitable.

OTHER INSECTS. Many kinds of butterflies, bees, and flies are active in visiting flowers, and in doing so they act as agents in the distribution of pollen. In onion breeding, species of flies have been used successfully for pollination. The onion plants to be crossed are grown in wire cages, and the flies are introduced in the cages at the proper time.

Wind. Some species of plants are pollinated by wind. The date palm, pecan, walnut, filbert, corn, spinach, and beet are examples of plants of this class. These plants do not have the conspicuous flower parts that characterize those pollinated by insects. The pollen produced by them is very fine, light, dry, and easily borne by the wind.

Investigations have shown that pecan pollen may be carried 3,000 feet by wind, and it is generally believed that it may be blown much farther. Rain, dew, and fog inhibit the distribution of pollen by wind. It has been found that no shedding of pollen occurs in the pecan when the relative humidity of the air is above 85 per cent. Prolonged periods of rainy or extremely humid weather thus tend to reduce fruitfulness in wind-pollinated plants. Such unfavorable weather during the pollination period may account, in some degree, for the uncertainty of crops of wind-pollinated species in sections where such conditions are likely to occur. With some plants, both wind and insects are responsible for pollination. Such is the case with grapes.

Hand-pollination. The pollination of flowers by hand is a rather common practice in experimental work. It is also followed to a limited extent in commercial production of some crops, such as apple, where for one reason or another insect pollinators are not present in sufficient numbers to ensure a good set of fruit. Such controlled pollination eliminates the necessity for fruit thinning to prevent overbearing.

Fertilization. The stigmatic surface of the pistillate flower, when the latter is receptive, usually has a sticky or viscid fluid upon it. The pollen grains, falling on this favorable medium germinate in a comparatively short time, and a tube grows out from one of the pores of the grain. This tube, known as the *pollen tube,* penetrates the style of the flower, supposedly by dissolving its way beween the cells, and finally penetrates the ovule. As it grows, the nucleus of the pollen grain, which has divided into two parts, follows down the tube. When the tube finally emerges into the ovule, the two nuclei are discharged into the embryo sac. The time required for this process varies, com-

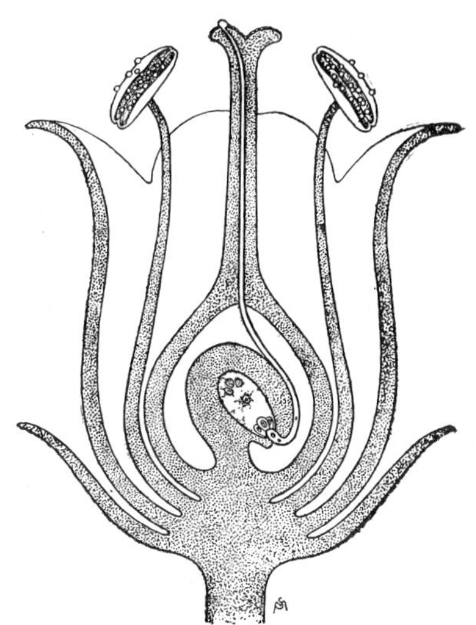

Fig. 22. Diagram to show the general anatomy of a complete flower and the process of fertilization.

monly, from a few hours to as much as 5 days; in the pecan it requires from 2 to 3 weeks, and in the overcup oak it requires 1 year. In the meantime, within the ovule, certain changes have taken place that finally result in eight nuclei, arranged within the embryo sac. These consist of the egg cell and two synergids at one end, three antipodal nuclei at the other end, and two polar nuclei near the center. As the two generative nuclei are discharged from the pollen tube, one of them fuses with the egg cell, producing the embryo of the seed. The other generative nucleus usually fuses with the two polar nuclei, to produce the endosperm. This complete process is known as *double fertilization.* In the first growth of the embryo after fertilization it becomes differentiated into certain regions. It is held in place by a suspensor, which may consist of only three or four cells, or it may be only one cell wide and very long. Next in order of development are the radicle, which is the root in the young plant; the hypocotyl, which is the lower stem; the epicotyl, which is the stem between the cotyledons and the first true leaves; the cotyledons, or seed leaves; and the plumule, or growing point, of the young plant. As the seed develops and matures, food materials in concentrated form are deposited within the seed coat. These consist of carbohydrates, proteins, and fats in varying proportions, and different kinds predominate in the different seeds. In some seeds stored food is largely confined to cells of the embryo; in others it is in the endosperm adjacent to the embryo.

Seeds. Seeds are extremely variable in size, shape, color, and length of time during which they remain viable. Certain plants, of which celery is an example, produce seeds that are as small as grains of sand; and others, such as the begonia, produce seeds that are even smaller. Avocado seeds are often as much as 2 inches in diameter. In shape and texture seeds may be rounded, angled, smooth, irregular, thin, or flat. With all these variations, true seeds have three essential parts in common, which become their distinguishing characteristics: (1) The embryo is the most important component of a seed. It is the living plant developed from the fertilized egg cell, and its growth has been restricted by the maturity of the seed. Its parts are the radicle, plumule, and one or two cotyledons. (2) Stored food is another component of a seed. It is deposited in the seed while it is still on the mother plant. This reserve food may be contained largely in the embryo, as in the peach and bean; or in the endosperm which closely encloses the embryo of certain kinds of seeds. Seeds of this latter kind are produced by the onion, for example. (3) The testa is the outside covering forming the protective coat of a seed. It is formed normally from the two integuments of the ovule, or in some seed from a single integument.

Multiple Seedlings. Seeds of some plants are contained in fruits, with or without adhering parts. Such fruits may contain one or more than one seed and produce a corresponding number of seedlings. The so-called lettuce "seed" is a fruit that contains only one seed and it produces only one seedling when it germinates. The so-called "seed" of the beet is a fruit that contains several seeds and hence when planted, it produces several seedlings.

Some true seeds contain more than one embryo. Such seeds upon germination produce more than one seedling. This condition is known as *polyembryony*. It may result from several causes including (1) the development and fertilization of more than one egg cell in the embryo sac, (2) the occasional successful fertilization of synergids, and (3) the development of vegetative embryos in the nucellus of the ovule. In the latter case, the fertilized egg produces one embryo which has heritable factors from both parents. In addition to this embryo, a number of vegetative embryos are produced from the nucellus of the ovule. These produce seedlings that have the same characteristics as the mother plant. The vegetative embryos and the seedlings that grow from them are *apogamic*. Apogamy is fairly common among plants. It has been shown that at least one species of the apple produces apogamic embryos, and it has been observed frequently in citrus. In one test Imperial grapefruit produced an average of four seedlings per seed and Willowleaf Mandarin produced over six seedlings per seed.

Fruit. From the botanical viewpoint, the fruit is the matured ovary with its seeds and other parts of the flower that are intimately associated with it at maturity. The fruit then usually consists of seeds, which are developed from ovules; pericarp, which develops from the ovary; and perhaps other parts, that will be discussed later.

Simple Fruits. These fruits develop from a single enlarged ovary to which other parts may or may not be attached. If the ovary and later the pericarp is borne *upon* the receptacle, as in the grape, the flower and fruit are superior or *hypogynous;* if they are *enclosed* and *surrounded* by the adhering receptacle, as in the apple and muskmelon, the flower and fruit are inferior or *epigynous;* if they are borne within a cup-shaped receptacle, as in the peach and plum, the flower and fruit are *perigynous*.

Simple fruits may develop from an ovary of one carpel, as in the peach and bean; or more, as in the grape with two, the cucumber with three, okra with four, the apple with five, and the orange and common tomato with many. The pericarp of simple fruits may be fleshy at maturity, as in the tomato and grape; or dry as in walnuts, pecans, and filberts.

Aggregate Fruits. An aggregate fruit is derived from a single flower having a large number of pistils. The structure of such a fruit is that of a single receptacle upon which are massed a large number of so-called "fruitlets," each of which is like a small fruit. In the dewberry, blackberry, and raspberry, the individual fruits are small drupes; in the dewberry and blackberry these drupelets adhere to the receptacle when it is detached from the stem, while in the raspberry, the receptacle adheres to the stem and the fruit separates as a hollow cup. In the strawberry, the small fruits are achenes or dry fruits and the receptacle is the chief edible portion.

Multiple Fruits. The multiple fruit is developed from the ovaries of many separate flowers. These are closely clustered together on one peduncle and they may be sessile or borne on pedicles. Good examples of these fruits are the mulberry, pineapple, and fig. The individual flowers in the pistillate inflorescence of the mulberry are crowded together closely on the axis. Each flower possesses a single, one-carpel ovary, which develops into a nutlet enclosed by the thickened, juicy calyx lobes. These separate fruits become crowded together as they develop to form the mulberry "fruit."

The flowers of the fig are borne on the inner wall of an enlarged fleshy hollow peduncle sometimes known as "receptacle," staminate and pistillate flowers occurring in some types of figs and only pistillate flowers in other types. The pistillate flowers have single, one-carpel ovaries, developing into nutlets imbedded in the succulent flesh of the true receptacle inside the hollow peduncle. The edible portion consists of these flowers, their short pedicels, and the fleshy peduncle. The pineapple has an elongated central axis on which are borne numerous sessile flowers. The fleshy bases and ovaries of these flowers are fused with sepals to form the edible part of the pineapple.

Accessory Fruits. Those fruits in which a part of the ripened fruit has developed from parts other than the ovary and ovule are known as accessory fruits. The apple is a simple fruit but also accessory because a large part of the flesh is derived from the receptacle. In general, fruits that develop from inferior (epigynous) ovaries are accessory. Accessory fruits, however, develop also from hypogynous ovaries, as represented by the strawberry, dewberry, and blackberry. In the strawberry the receptacle comprises practically the entire edible portion and in the blackberry and dewberry a part of it. All these are, hence, accessory, aggregate fruits. The mulberry is a multiple, accessory fruit in that the "fruit" consists of peduncle and fleshy

FIG. 23. Multiple fruits of fig, borne in axils of leaves.

bracts in addition to that which develops from the pistil. The fig fruit is also a multiple, accessory fruit.

Seedless Fruit. Some plants commonly produce fruits that contain no seeds. The conditions responsible for seedlessness may be conveniently discussed under two headings:

1. *Parthenocarpy.* Fruits that set and mature without being fertilized are designated as parthenocarpic.

a. When plants set and mature fruit without pollination, the condition is known as *vegetative parthenocarpy*. The banana,

Japanese persimmon, English varieties of cucumbers, orange, grapefruit, and some varieties of figs are examples of plants characterized by this condition.

b. In some cases, the stimulation of pollination on the ovarian tissue is essential, although fertilization of the ovule does not follow. This condition is known as *stimulative parthenocarpy*. Examples are found in the pear, Jerusalem cherry, Thompson Seedless grape, and some varieties of squash. Thompson Seedless grapes may be rendered seedless also, by embryo abortion, discussed in a later paragraph.

c. If fruits are capable of developing into seedless fruit if not fertilized, or into seeded fruits when fertilized, they are said to be *facultatively parthenocarpic*. Certain varieties of Japanese persimmons, figs, cucumbers, and peaches are examples of fruits of this class. The J. H. Hale peach produces some fruits which are seedless. They are invariably smaller and later in maturity than normal fruit.

d. If fruits will not develop seeds under any circumstances even when pollinated, the parthenocarpy is said to be *obligatory*. The Thompson Seedless grape is an example.

Fertilization of parthenocarpic fruits might be objectionable by causing the fruit to develop seed instead of being seedless, thereby causing the mother plant to be less fruitful. In the case of cucumber varieties that are capable of maturing fruit parthenocarpically, flowers that are pollinated and fertilized produce seed-bearing fruits which are usually different in shape from seedless fruits. They tend to be angular in cross section instead of cylindrical. Cucumber vines of varieties that produce parthenocarpic fruits are very prolific of seedless fruits; if pollinated, the cucumbers have seed and the vines are less prolific.

The Hachiya variety of Japanese persimmon will set fruit readily without pollination, in which event fruits are seedless. In mixed plantings where pollination is likely to occur, seeded fruits are frequently produced. These have black, discolored areas immediately surrounding some of the seeds and are considered to be inferior in quality to the seedless fruits.

2. *Embryo Abortion or Killing.* Seedlessness may be due to causes other than parthenocarpy. As a result of the fertilization process, an embryo is formed. If its growth is not arrested, it forms a seed. In some fruits the growth of the embryo may be

stopped by internal or external factors, and yet the fruit will continue to develop. Embryos are often killed by cold or other conditions that do not kill the ovary.

If the fruit has developed far enough so that its growth will continue without the stimulus of the developing seeds, a seedless fruit is formed. However, as a rule, death of the embryo results in premature shedding of fruits. If fertilization takes place in Thompson Seedless grapes, the fruits when mature are seedless because of embryo abortion. This phenomenon has been observed, also, to result in seedless fruits in plum and cherry, though such fruits have no special merit because of the presence of the bony endocarp.

Factors Influencing Fruit Setting. In general it is necessary that flowers be fertilized in order to set fruit. Parthenocarpic fruits are an exception. There are several factors that may operate to discourage or prevent fruitfulness; these are therefore of vital concern to the fruitgrowers.

Defective Flower Parts. Incomplete development of pistils or stamens may be a cause of sterility within a species. Many species of native plums and other fruits regularly produce some flowers with defective pistils. These are incapable of being fertilized and they drop from the tree at the time of petal fall.

Muscadine grapes produce some flowers in which the pistils are normal, but the stamens are abnormal. The pollen produced by these stamens is shriveled instead of normally plump and smooth. Such pollen is not viable. The J. H. Hale peach produces flowers with normal pistils, but the anthers are abortive and therefore are pollen-sterile. Some varieties of hybrid tomaties are known to produce normal amounts of pollen, but its development is not complete and it is sterile at the time of dehiscence. The Italian Red onion does not normally produce viable pollen. It is obvious that self-pollination cannot occur in such flowers in which either the pistil or anthers is defective.

Incompatibility. Pollination does not always ensure fruit setting. In many fruits incompatability exists between pollen and pistils of the same plant. The pollen is viable, but it is not capable of effecting fertilization of the pistil of the same plants. Such conditions result in self-sterility. The pistils of such plants, when pollinated with the proper pollen, will become fertilized and develop normally. Several of the important commercial varieties of apples are known to be self-incompatible, or partially so. Some of the leading pear varieties, including Bartlett, Kieffer, Bosc, and Anjou, are regarded as commercially self-sterile, at least in certain localities. Many of the varieties of plums, such as Abundance, Wickson, Burbank, and Bruce, are self-sterile. The same is true for certain varieties of almond, cherry, and blackberry. For all of the varieties that are known to be self-sterile at least one other variety known to be an effective pollenizer should be planted.

When incompatibility exists between pollen and pistil of different varieties or species, they are said to be *intersterile*. Each of three leading varieties of sweet cherries (Bing, Lambert, and Napoleon) is self-sterile and the three are intersterile. Hence, mixed plantings of them will not produce fruit unless the trees are within range of some other variety, such as Black Tartarian, that is interfertile with them. The Bartlett and Seckel varieties of pears are each commercially self-sterile; and the two are intersterile and hence are not fruitful when planted alone or together without an effective pollenizer, such as Bosc.

Dichogamy. Plants frequently exhibit a difference in the time of pollen shed and the time of pistil receptivity. The period when pollen is shed may not coincide with the time when the pistil is receptive. If the two periods are entirely distinct, the condition is known as *complete dichogamy;* if some overlapping occurs, it is *incomplete dichogamy.* For example, a variety with pollen shedding from April 11 to April 18 and pistils becoming receptive from April 20 to April 28 would represent a condition of complete dichogamy. If pollen is shed before pistils are receptive, the plant is said to be *protandrous;* if pistils are receptive first, *protogynous.* Isolated plants of a dichogamous variety are normally unfruitful. Plants that are incompletely dichogamous, if isolated, will set fruit only during the period of overlapping, and the quantity of fruit set is determined largely by the length of the period during which pollen shedding and pistil receptivity coincide. It should be kept in mind that dichogamy also results in self-sterility or intersterility, not as a result of incompatibility but merely through the operation of the time factor.

Data obtained in Texas showed dichogamy to be a factor limiting the fruitfulness of pecans. Many of the leading varieties have been observed to be protogynous in most years, and in only a few cases was there sufficient overlapping in blooming to allow self-pollination. A few varieties, on the other hand, were found always to be protandrous and produced pollen each year in time to pollinate any variety that had been under observation. At the time that the pistils of early pollenizers became receptive, pollen from other varieties was available.

Dioecious Plants. Plants that bear only staminate flowers never produce any fruit, since no ovary capable of developing into a fruit is borne; isolated plants that bear only pistillate flowers are seldom fruitful unless some special provision is made to ensure effective pollination. The Smyrna fig, a dioecious species that bears only pistillate flowers, was not fruitful when first introduced into the United States because pollen that was necessary for fruit setting was not provided. In pollination of this fruit, Capri figs containing insects are placed in wire baskets and hung in the Smyrna trees. These are replaced at intervals of 2 or 3 days in order to ensure a continuous supply of viable pollen. The Capri figs are grown in protected locations to prevent possible killing by cold weather of the overwintering crop of figs, which contain the fig wasps. A similar practice of planting is followed with the male date palm. With some dioecious species, such as the Muscadine grape and pistachio, the staminate plants are interplanted to provide proper pollination.

Environmental and Nutritive Factors. In addition to the factors mentioned above, fruit trees may fail to set fruit properly, owing either to frost and other climatic factors or to internal nutritive conditions. The destructiveness of frost and cold weather is often quite apparent where warm winter weather causes early blossoming of peach, plum, and other fruits. In addition to the damage due to outright killing of the pistils, cool weather may slow pollen-tube growth so that the ovule does not become fertilized before an abscission layer forms and causes dropping of the flowers.

High humidity may prevent pollen shed, as in the pecan. Low humidity may shorten the period of receptivity of the pistil, as in the tomato; unfavorable weather may discourage bees or other insect pollinators and thus prevent pollination.

Unfavorable nutritive conditions within the tree often result in its failure to set fruit. Partly developed fruits may drop off, owing to the plant's inability to furnish the necessary nourishment required in growth of vegetative and flower parts. The nutritive factor may often explain why certain varieties bear a heavy crop one year and a very light crop during the next season. The explanation may be in the fact that undernourished or weak trees often produce defective pistils, fail to develop viable pollen, or fail to develop any flower buds at all. It may determine, also, which of the branches will bear fruit; evidence indicates that those which are more vigorous or are in a more favored position with respect to a nutrient supply are more likely to bear fruit. This is especially noticeable during light-crop years.

Practices That Increase Fruit Setting. It is suggested in previous paragraphs that some plants are not fruitful because viable pollen is not transferred to the receptive stigma at the proper time for effective pollination. Several special practices are followed by orchardists in overcoming this difficulty:

1. Growers who have discovered an immediate need of pollen may provide it by distributing through the orchard blossoming branches of a variety that produces effective pollen during the time when the pollen is needed. Large branches, from 1 to $1\frac{1}{2}$ inches or more in diameter, are preferred. The pollen of flowers of such branches matures and is available for distribution by wind or insects for several days. The branches are placed in vessels of water and are replaced by fresh ones as the flowers wither.

2. A permanent source of such pollen may be provided by grafting into each tree a scion of a good pollenizer. As the scion grows, it becomes a part of the tree and provides pollen for future years. One pollenizing branch on each tree is adequate. The chances of having pollen available when needed are increased by grafting a relatively early blooming variety on one tree and a later one on the adjacent tree.

3. Planting pollenizing trees in the orchard at intervals may also be used to provide a permanent source of pollen. A rather customary practice is to plant one pollenizer to each 10 trees. In many cases, one or more varieties which will pollinate the commercial variety and produce good-quality fruit can be interplanted.

4. The location of bees or other pollinators in the orchard for pollinating purposes is a rather common practice in some sections. One colony to the acre is considered adequate. It is not

necessary to distribute the hives singly throughout the orchard. They may be placed in groups of eight or ten on a site exposed to the sun and protected from the wind as far as possible. The bees should be brought into the orchard just as the first bloom opens so that they will be encouraged to begin work nearby instead of elsewhere.

QUESTIONS

1. What are the three essential parts of a seed? What is the important function of each part?

2. Name the parts of a complete flower. What is a perfect flower? An imperfect flower? A monoecious species? A dioecious species?

3. Define pollination, self-pollination, and cross-pollination.

4. What are the principal agents of pollination?

5. What weather conditions discourage pollination by wind?

6. What is meant by dichogamy?

7. Tell what is meant by polyembryony or apogamy. Of what significance is it?

8. What is an inflorescence?

9. What is the essential difference between a determinate and an indeterminate type of inflorescence? Between a spike and a raceme? Distinguish between sessile and pediceled flowers.

10. What is the distinction between a fruit and a seed? Between simple, aggregate, and multiple fruits?

11. Distinguish between fruits with fleshy pericarps and those with dry pericarps. Give examples.

12. Give examples of accessory fruits. What tissues comprise the edible part of each?

SUGGESTED REFERENCES

Baldwin, H. I.: "Forest Tree Seed of the North American Temperate Regions," Chronica Botanica Co., 1942.

Crocker, William, and Lela V. Barton: "Physiology of Seeds," Chronica Botanica Co., 1953.

Frost, H. B.: Polyembryony, Heterozygosis and Chimeras in Citrus, *Hilgardia,* **7:**625–642, 1933.

Griggs, W. H.: Pollination Requirements of Fruits and Nuts, *Calif. Agr. Expt. Sta. Circ.* 424, 1953.

Hedrick, U. P.: "Systematic Pomology," The Macmillan Company, 1925.

Pool, R. J.: "Flowers and Flowering Plants," McGraw-Hill Book Company, Inc., 1929.

Snyder, John C.: The Pollination of Tree Fruits and Nuts, *Wash. Agr. Ext. Serv. Bull.* 342, 1946.

CHAPTER 4

Special Plant-growing Equipment

Special equipment is used in the growing of many horticultural plants. This special equipment is used to start plants at seasons when outside conditions are unfavorable, to grow plants to maturity at off seasons of the year, and for the propagation, by seed or vegetative methods, of plants that require special treatment.

Types of Special Equipment. There are several different types of kinds of special equipment. The kind or species of plant to be grown, the length of time the equipment is needed during a season, initial cost, operating expenses, and other similar factors are considered in deciding upon equipment to be used.

Forcing Hills and Plant Protectors. Certain structures are designed to cover the individual plants in the field. They are known as *forcing hills* or *plant protectors*.

Types. Several types are used. Formerly, one made in the form of a box, usually 12 inches square and 12 inches high with a pane of glass for cover, was used extensively. The initial cost, cost of storage, breakage, and labor required to place them over the plants and remove them have all tended to discourage their use.

Other types of plant protectors, however, are used for growing plants in the open. Small conical covers made of translucent paper or plastic are used commonly. Some plastic covers are flexible and are designed for using once only; others are rigid and may be used repeatedly.

Uses. Plant protectors and forcing hills are used to protect plants from untimely cold weather and from damage by wind. They are also used to increase the soil temperature to a degree which is favorable for the germination of seed. Workers in Arkansas have shown that muskmelon seed planted early in the spring germinate quicker when plant protectors are used, because of the higher prevailing soil temperature. The plants which got an early start ultimately produced marketable melons at a slightly earlier date than plants in locations where no covers were provided. Forcing hills and plant protectors are used only for crops that produce a heavy yield of a valuable product from an individual plant. The tomato and muskmelon are examples of such plants. On the contrary, it would be impracticable to

Fig. 24. Conical paper covers used to protect tender plants in the field.

use forcing hills for carrots or radishes because the unit return from an individual plant of such crops is too small to justify the expense for labor and material.

Cold Frames and Hotbeds. Cold frames are designed primarily to protect plants from cold without the use of artificial heat. Hotbeds differ from cold frames in that they are provided with artificial heat.

Uses. Cold frames and hotbeds are used widely in the starting of vegetable crops, and to a lesser extent for cuttings. Cold frames are used primarily in protecting plants against a few degrees of cold, usually in early spring. They are also useful in providing protection against wind and excessive rainfall, and in the hardening of plants prior to transplanting to the field, a practice that is discussed in the chapter on transplanting. In some places, certain crops are started in cold frames and, when the weather permits, the frames are removed and the crops continue to grow under field conditions. Plants may be grown in hotbeds at seasons when it would be too cold for them in cold frames; hence the season of profitable use of a hotbed is much longer. Oftentimes, young plants are started in late winter in a hotbed and later as the weather becomes milder they are transplanted to the cold frame. After a period of growth there, they are finally moved to the field when outside weather conditions have become favorable.

Fig. 25. Hotbed, standard sash covers, and mechanical conveyer to facilitate handling of sash.

Construction. Cold frames and hotbeds are constructed in the same general manner. They are usually made of wood or concrete. When wood is used, the structures can easily be made so that they are movable. This makes it possible to set them up at different places each year and to store them during off seasons. Insulating the walls, particularly those made of wood, makes them more effective in retaining heat and providing protection. This is commonly done by lining inside walls with heavy paper or by banking soil against the outside of the walls. The standard width of cold frames and hotbeds is 6 feet; the length is variable, depending upon the space needed. Cold frames and hotbeds should be located on the south side of a building or other barrier which will provide protection from north winds. The lengthwise direction should be from east to west. The north wall of the structure should be 6 inches higher than the south wall. This facilitates shedding of water when the frame is covered. It provides better exposures to sunlight in late winter and early spring and also provides some protection from north winds. The bed or floor of the cold frame or hotbed should be level to facilitate uniform watering, and it should be even with, or slightly above, the surrounding ground level to ensure good drainage. When concrete is used to make a permanent structure, the walls usually extend well into the soil and therefore special

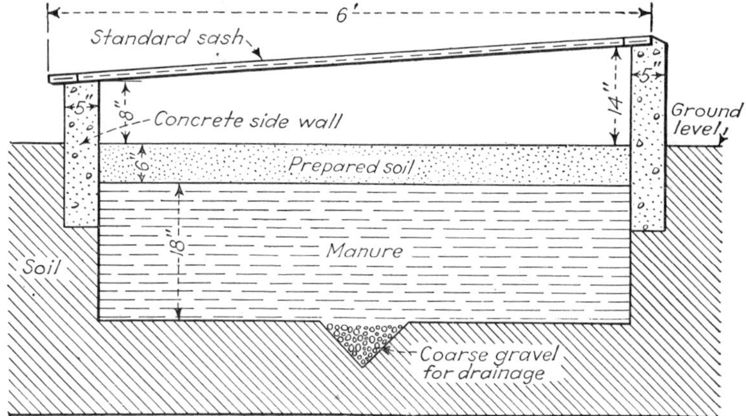

Fig. 26. Cross-section drawing of a manure-heated hotbed.

provision should be made to provide adequate drainage by the use of a sand or gravel fill and tile drains.

Covers. Normally, covers of some kind are used for cold frames and hotbeds. The most satisfactory cover is the standard sash. It is 3 feet wide and 6 feet long. Glass panes are imbedded in the frame and glazed to provide waterproof and airtight protection. In use the sash is placed lengthwise across the cold frame or hotbed. The standard sash is expensive; yet it is a satisfactory cover. A frame covered with glass permits the absorption of heat from the sun on clear days and it enables the bed to retain it during the night and during cold periods; it is possible in this way to provide temperatures that are more uniformly favorable for plant growth than would be the case if the frames were not so covered. Various other materials are used as covers for cold frames. Screen wire imbedded in a transparent material similar to cellophane makes a satisfactory cover. This material is usually tacked on frames of dimensions that permit of convenient handling. Different grades and weights of cloth that range from heavy duck to light domestic are also used. The untreated cloth may be used, but treating the material with hot linseed oil or melted paraffin increases its durability, makes it more nearly waterproof and airtight, and renders it more effective in protecting the frame during unfavorable weather.

Methods of Heating Hotbeds. Heating of hotbeds is accomplished in four principal ways.

HOT WATER OR STEAM. Where hotbeds adjoin a greenhouse that is heated by steam or hot water, the heating pipes may be extended into the beds also. Other provisions are sometimes made for steam or hot water. The pipes are usually placed about 5 or 6 inches below the seedbed surface. Where it is desired to protect plants against an occasional late frost or freeze, and where it is desirable to warm the air, but not necessary to warm the soil, the pipes may be suspended along the inside walls at about the level of the seedbed. Hotbeds heated with steam or hot water are very satisfactory because the temperature can be regulated accurately.

ORGANIC MATTER. The heat liberated in the decomposition of organic matter can be used as a source of heat for hotbeds. Animal manures are used commonly and fresh manure from grain-fed horses is considered best. Hay, straw, and cornstalks are also used, though the heat produced by these is much less. The hotbed is excavated to a depth of from 18 to 30 inches. The manure or other organic material is packed well into this basin, especially around the edges and in the corners. When the required amount has been added, a layer of good soil, 4 to 6 inches deep, is spread smoothly over the top. This constitutes the seedbed and its surface should be slightly higher than the level of the surrounding ground. When moisture is added, heat is produced by organic material and the seedbed above absorbs some of the heat. The greatest heating effect is at the beginning of the period, and the temperature gradually subsides. Hence, this type of hotbed is more satisfactory for use in the spring than in the fall. If manure, or other organic material, is available locally, the chief expense of providing heat is the labor necessary to put the bed in operation.

FLUE HEAT. By another method, hotbeds are heated by flues. In the construction of such beds, a firebox is located at one end and tile flues extend from the firebox lengthwise of the bed to an outlet at the opposite end. Two lines of flues, properly spaced, give a more uniform distribution of heat than if only one line is used. Soil is placed over the flues to provide the planting bed. Hot gas and smoke from the firebox, passing under the bed, create the heating effect. Cheap fuel is essential for the practical operation of a flue-heated hotbed. Wood has been used more commonly than any other fuel, but high labor costs are making it more expensive. Careful and regular attention is required to provide uniform heat; hence the labor cost of operation is high. They are inconvenient to operate, particularly when it is necessary to provide heat day and night for a prolonged period.

ELECTRICITY. As electricity becomes more generally available, it is being used increasingly in the heating of hotbeds. Light bulbs, mounted on suitable panels, and suspended in the air within the hotbed, may be kept burning to keep the air temperature above the danger point during short cold periods. Several low-watt-power globes distributed over the entire area to be heated are preferred to a smaller number of high-watt-power globes. In addition to the heating effect, light bulbs provide supplemental light which is advantageous in some cases. Special lead- and plastic-covered heating cables are now available for heating hotbed soil. The cable is laid back and forth across the bed at intervals of 6 to 8 inches. Soil is added to make a seed-bed 4 to 6 inches deep over it. A thermostat may be used to control the temperature at which the electric current will cut off and on. The soil temperature to be provided varies with the different kinds of plants to be grown in the hotbed. For tomato and sweetpotato the thermostat is set so that the current will be cut off if the soil temperature rises above 85°F. and will come on again if the temperature drops below 75°F. When the cable has been installed with a thermostat, a favorable soil temperature is provided automatically and the labor cost for operation is reduced to a minimum. The amount of electricity required, and hence the cost for heating hotbeds, depends principally upon (1) the temperature required, (2) the amount of cold weather which prevails, and (3) the type of hotbed and covers used.

With sweetpotatoes in Texas, 13 kilowatthours of electricity were required to produce 1,000 plants when the bed was covered with standard sash, but 23 were required when a cloth covering was used. Beds banked with soil (as insulation) required 27 kilowatthours for each 1,000 plants, while 51 were

Fig. 27. Flexible electric heating cable laid in hotbed, to be covered with soil. (*Courtesy of P. T. Montfort, Agricultural Engineering Department, College Station, Tex.*)

required for the beds without insulation. The plants were 5 days earlier in the beds covered with standard sash and in the insulated beds than in check beds. For several locations in Texas the electricity for heat to produce sweetpotato slips ranged from 1.7 kilowatthours per 1,000 plants where the weather was mild, to 26.8 in locations where the weather was cooler. For tomatoes, the requirements for 1,000 plants ranged from 7 to 13 kilowatthours at different locations, depending upon the amount of cold weather that prevailed during the period of operation. Oftentimes costs are calculated upon the electricity required to provide heat for the area covered by one standard sash. Thus tests in Washington and Pennsylvania show that the cost during a certain period was almost 3 cents per sash area per week; while under different conditions, in Maryland, the comparative cost was only about 1.5 cents per sash per week.

Greenhouses. The glass-covered house represents an improvement over the cold frame and hotbed and is superior to the other forcing structures for starting plants. Temperature may be controlled more accurately, ventilation regulated more perfectly, and arrangements more convenient for work provided.

Fig. 28. View of a good type of greenhouse. Note curved eaves.

Greenhouses are constructed in many sizes and types. The cost need not necessarily be exorbitant. Greenhouses are sometimes made by providing low walls and using standard sash for roof.

Types. In design there are three distinct types of greenhouses:

1. The *lean-to* type is constructed on the side of a building, usually the south side. The roof slants in one direction. Its length should extend in an east and west direction. When so located, the roof will necessarily slope toward the south. This is desirable since it permits better exposure to the sunlight when it is needed most during winter and spring.

2. The *three-quarter-span* greenhouse is a type in which three-fourths of the surface of the roof slopes in one direction usually

toward the south; one-fourth slopes in the opposite direction. The ridge of the roof is hence off-center. The southern slope of the longer span permits favorable exposure to sunlight; the quarter-span makes it possible to ventilate more effectively. This type of greenhouse should extend lengthwise from east to west.

3. The *even-span* type of greenhouse is one in which the roof slopes evenly in two directions, the roof ridge being above the center of the greenhouse. The most uniform exposure of the greenhouse to sunlight is obtained if the lengthwise direction is north and south.

Construction. The greenhouse is designed to provide protection against cold and to permit exposure of plants to the maximum amount of sunlight. The sizes of the supporting framework members should be small to prevent shading, and the side walls should be low. Ventilation is an important factor in the management of greenhouses. The best ones have ventilators on each side of the roof ridge and also one or two on each side extending the length of the greenhouse.

Two general types of beds are used for the growing of plants. One type is the ground, or "solid" bed in which the plants are grown in specially prepared beds at ground level. The other type is the so-called raised bed, or bench type. These are more convenient for planting and working. They are usually made of concrete and, if well constructed and reinforced, will last indefinitely. For both ground and raised beds, good drainage is essential. This can be easily provided in raised beds, by sloping the floor to drainage outlets, spaced at proper intervals. It is customary to make such beds sufficiently deep so that a layer of coarse sand or gravel can be put in the bottom to provide good drainage. A bed 8 inches deep will permit a 2-inch base layer of sand or gravel and 6 inches of medium for plant growing.

Heating. Hot air, hot water, and steam are used for heating greenhouses. Steam and hot water are probably the most desirable, especially in sections where continuous heat is required. Burning gas or other fuel in the greenhouse is sometimes practiced where heating is required only at intervals. When this is done, ventilation is essential to reduce damage from fumes. Automatic gas heaters, provided with blowers and thermostats, are in common use in greenhouses, and they are fairly satisfactory.

Slatted Frames, or Semishades. Many horticultural plants are tender to heat, and require protection from the heat for the best growth. The strawberry plant, for example, does not withstand extreme heat. Cabbage plants for a fall crop in the South must often be started in late summer when the temperature is unfavorably high. Many ornamental plants can be grown best where shade is provided. Partial shade facilitates the growing of such plants in regions where a high temperature prevails during all, or a part, of the growing season.

Fig. 29. Slatted frame constructed so as to be movable.

Various types of slatted frames or semishades are used to provide partial shade for tender plants. A satisfactory slatted frame can be made by spacing 2-inch strips horizontally at intervals of 2 inches over a framework of convenient height. One-half of the area is thus covered. The sides may also be stripped, depending upon the protection needed for the plants to be grown. The frames may be made in permanent locations, or of a size and design that permits them to be easily moved from one location to another. There is an advantage of constructing them so that they can be moved. It permits the seedbed to be constructed in a different place each succeeding year, and the placing of the slatted frame over it after the soil has been prepared and planted.

Shading can also be provided by using building paper supported between coarse-mesh wire. Roll picket fence supported by a suitable framework can be used to provide a suitable semishade. The height of such shades is adjusted to permit desired light from the side.

Since all of these types of semishades are designed to protect plants from excessive heat instead of cold, they are used principally in hot climates.

Propagating Beds. Outside beds are very useful in certain types of propagation. Such beds can easily be made by using concrete tile blocks for the border. Six feet is a convenient width for them, and they can be made of any desired length. When filled with sandy loam soil, such beds are suitable for the growing of seedling plants to be used for lining-out stock or for other purposes; the rooting of certain types of cuttings, principally hardwood; and for the planting of whip grafts made indoors.

Factors of Management in Plant Growing. The use of good equipment does not in itself carry with it the assurance of success. Successful operation depends also upon good management. This applies equally to the hotbed, cold frame, greenhouse, and semishade. Management involves the primary problems of heating, watering, ventilation, and control of insect and disease pests. It involves also the use of good soil and the practice of certain rules of technique with respect to the actual details of operation.

Soil. The use of the appropriate type of soil or media is one of the most important factors of good management. Sand and sandy loam soil are frequently provided for beds where cuttings are to be grown. Other materials are also used. For the growing of seedling plants that are to be transplanted, it is important that good soil be used in the seedbed. The soil may determine whether the plants are stocky or spindling, vigorous or stunted, normally developed or excessively luxuriant. Soil likewise influences directly the vigor of plants that are to grow to maturity in the forcing structure. It should be fairly fertile, of good physical texture, well aerated, and relatively free from insects and disease organisms.

Soil Preparation. Specially prepared soil should be provided for seedbeds. The most valuable source is from compost beds. Compost is made by stacking alternate layers, about 4 inches thick, of good loam soil and barnyard manure until the pile is

4 or 5 feet high. It may be of any convenient width and length. The top should be concave so as to hold water that will soak into the layers and encourage decomposition. The compost pile is prepared from 1 to 2 years in advance of the time it is to be used. In the meantime, however, it should be spaded vertically and restacked so as to blend the loam and manure. This should be done every 3 or 4 months. A mixture of compost, sand, and loam makes an ideal soil for use in seedbeds or for potting. Acid peat is oftentimes added to ensure friability and good aeration; and fertilizer may be added to increase fertility. One blend that

Fig. 30. Interior of greenhouse used in propagation work. Plants in flats in foreground. Hardwood cuttings in bed in center.

has been used satisfactorily by the authors is two parts compost, 4 parts loam soil, 2 parts builder's sand, 1 part acid peat, and 1½ pounds of 10-20-10 commercial fertilizer per cubic yard of soil.

Soil Sterilization. Insects and diseases in the seedbed soil may cause losses in several different ways. The seedling plants may be killed while they are still in the bed, or they may become affected there, though trouble does not develop until the plants approach maturity in the field. The seedbed may be a source of infestation whereby insects and diseases are carried to the field on the roots and adhering soil of the plant. Such pests once introduced may become established permanently and discourage the growth of crops in future years.

HEAT. Heat was used first in sterilizing soils and it is still a popular treatment. Formerly, a fire was burned over the seedbed area to kill all organic life in the top layer of soil. Steam is used commonly as a source of heat to sterilize soil in closed containers. Boxes with insulated walls are constructed to hold a given quantity of soil, usually about a cubic yard. Steam is admitted through perforated pipes or open channels until the

FIG. 31. Soil-sterilizing unit, showing electrical heating elements.

soil throughout the box has attained a temperature of about 175°F. The soil is usually allowed to stand from 2 to 4 hours after the steam is cut off. This increases the effectiveness of the heat in destroying objectionable organisms. Outside beds are also sterilized with steam, by the use of metal pans as covers to retain the heat. Electricity is used effectively in heating soil, by a process called *soil pasteurization*. The soil in a closed container equipped with electrically heated flanges is heated to a temperature of from 150 to 160°F., and the temperature is maintained for 3 or 4 hours.

Better plant growth will be obtained if the soil, sterilized with heat, is stored for 3 or 4 weeks before it is used. This is to allow the beneficial organisms to build up in the soil so that the supply of nitrates will be replenished. In the meantime, it should be protected against contamination or reinfestation with objectionable organisms.

CHEMICALS. Certain chemicals are also effective when used to sterilize soils. Formaldehyde dust is effective in controlling most fungus diseases when mixed at the rate of 8 ounces of 6 per cent dust to 1 bushel of soil. Nematodes are killed by using 16 ounces

FIG. 32. Sixty-foot metal pan used in sterilizing field beds with steam.

of dust for a bushel of soil. Liquid formaldehyde solution, prepared by adding 1 pint of 40 per cent formalin to 30 gallons of water, applied at the rate of $\frac{1}{2}$ gallon per square foot, and a dilute solution of mercuric chloride (0.1 per cent) are also effective soil disinfectants. Zinc oxide spread over the seedbed surface prevents the spread of certain diseases. A red oxide of copper spray is effective in controlling damping-off fungi above ground on seedlings, and as it soaks into the soil, the disease is inhibited below the surface.

Other chemicals are used for the treatment of soil in place, without removing it from the bed. Chloropicrin is an effective soil fumigant for the control of soil pests. It is a poisonous gas, careful precautions must be exercised in its use, and special equipment is required for application. For those reasons other chemicals are used more commonly. Ethylene dibromide in solution with fuel oil applied to the soil at the rate of 9 cubic centimeters per square foot is effective in destroying nematodes, wireworms, white grubs, and other similar organisms that are objectionable in a propagating bed. Methyl bromide is likewise used generally to treat soils in special beds. An airtight cover, usually plastic, is provided for the beds. The rate of application is 1 pound for 100 square feet of bed area. The methyl bromide is contained in cans under pressure, and an inexpensive special applicator is required to introduce the gas under the plastic cover. The customary period allowed for treatment is 24 hours.

Watering. Uniform growth of plants depends on a uniform supply of soil moisture. Much of this must be applied artificially to plants in a special seedbed. Water that is free from objectionable salts should be used. The seedbed should be prepared so that the water will soak in uniformly and not run to one part of the bed. When plants are grown in pots, the soil should not completely fill the pot. Instead, each pot should have a slight basin at the top in order to facilitate watering. The size and depth of the basin should be uniform for all pots of a lot.

Temperature. Different kinds of plants require different temperatures for optimum growth. A suitable temperature for cabbage would be too cool for tomatoes. Plants of either species, when grown at a temperature slightly above optimum, are likely to make an excessively luxuriant growth. If the temperature is too low, they make limited growth and become dwarfed or stunted. Sudden changes in temperature should be avoided.

Lighting and Ventilation. Normal growth of plants depends on the proper amount of light. In the location and construction of the seedbed it should be remembered that light is an important factor. Shade created by an adjacent building, trees, high side walls, or crowding of plants tends to cause succulent and spindling growth. Such plants are undesirable from every standpoint.

Ventilation is a means of admitting fresh air to plants. It is the most effective way of regulating conditions of temperature and humidity. A combination of high temperature and high humidity is objectionable, because it encourages disease, especially damping-off. Plants that grow in an atmosphere that is very humid are likely to make a slender growth and become excessively succulent and tender. Such plants have a high water content, and they do not withstand transplanting well.

QUESTIONS

1. List types of special equipment used in growing horticultural plants.

2. What are the principal uses of forcing hills? They are used for certain kinds of crops, but not others. Explain.

3. What are advantages of hotbeds over cold frames? List covers used for hotbeds and cold frames.

5. What are the advantages and disadvantages of the different methods of heating hotbeds?

6. What are the reasons for using semishades?

7. What are the characteristics of good soil for starting young seedling plants?

8. What are the ways of sterilizing soil?

9. What precautions are made to insure adequate watering of plants?

SUGGESTED REFERENCES

Beachley, Kenneth G.: Combining Heat and Formaldehyde for Soil Treatment, *Penna. Agr. Expt. Sta. Bull.* 348, 1937.

Beattie, W. R.: Hotbeds and Cold Frames, *U.S. Dept. Agr. Bull.* 1743, 1935.

Crawford, Paul A.: Electric Hotbeds for Sweet Potato Slips, *Georgia Agr. Ext. Serv. Bull.* 533, 1950.

Horsfall, James G.: Pasteurizing Soil Electrically to Control Damping-off, *N.Y. (Geneva) Agr. Expt. Sta. Bull.* 651, 1935.

Hunter, F. M., and R. O. Monosmith: Propagating and Growing Plants with Electric Heat, *Miss. Agr. Ext. Serv. Bull.* 106, 1939.

Krove, Paul H.: The Reaction of Greenhouse Plants to Gas in the Atmosphere and Soil, *Mich. Agr. Expt. Sta. Bull.* 285, 1937.

Newhall, A. G., and W. T. Schroeder: New Flash-flame Soil Pasteurizer, *Cornell Univ. Agr. Expt. Sta. Bull.* 875, 1951.

CHAPTER 5

Methods of Propagation

Plant propagation is defined as the multiplication of plants. The propagation of horticultural plants includes the entire field of seed production and the growing of seedling plants; the production of bulbs and bulblike structures; the growing of plants on their own roots, by layerage and cuttage; and finally the production of varieties on special or selected rootstock, by budding and grafting. With certain crops the length of life of plants and their adaptability to the environment are determined by the method used to propagate the plants. This is particularly true and important for fruits, which are normally expected to grow and produce for a long span of years.

Origin of Varieties. Practically all the horticultural crops grown in the United States consist of named or standard varieties. The several methods of plant propagation are designed to perpetuate and reproduce these varieties. It is of interest, then, to study the ways by which these varieties have been developed. Briefly, they have come into existence in two general ways:

Seedlings. Many of our best varieties of fruits, vegetables, flowers, and ornamental plants have originated as seedlings.

CHANCE SEEDLINGS. Some varieties are discovered by chance or coincidence. In some cases, the seedling plants are grown for some other purpose or perhaps they are volunteer seedling plants. In other cases a large number of seeds are planted with the hope that one or more of the resulting seedlings will have sufficient merit to justify introduction as a named variety. The unusual and desirable qualities of the seedling may be observed at various stages of growth—either when it is young, or after years have elapsed and the plant has begun to bear. In any

62

event, if the resulting plant has good bearing quality, it can be used to start a standard variety. The Delicious apple, Elberta peach, Concord grape, and Thomas black walnut are examples of fruit varieties that have originated as chance seedlings. Likewise Hale's Best muskmelon is an example of a vegetable that has been introduced by selecting a chance seedling.

CONTROLLED POLLINATION. New or standard varieties may also be developed by controlled pollination. Pollen from known

FIG. 33. Granex, *lower center,* a hybrid of Grano, *upper right,* and Excel, *upper left,* onion varieties developed by plant breeding.

sources is used to produce seed which will produce seedlings from which selections may be made. Most of the current-day vegetable varieties have been developed in this way. Examples are Marglobe tomato and Congo watermelon. The Redhaven peach, Santa Rosa plum, and Ranger strawberry are examples of varieties of fruits that have been developed in this way.

Bud Mutations or "Sports." Quite frequently an individual limb occurs on a plant or tree which is different from all the other limbs on the same plant. The difference may be in color of foliage, size of leaf, length of internode, size, shape, color, or other characteristics of flower; or size, shape, color, or quality of fruit. Such unusual limbs are known as bud mutations or "sports." The actual mutation or sport may occur as a bud, though in most cases it would be quite difficult to identify until it had grown and produced a limb or twig.

Sports may be inferior or superior to the plant on which they occur. Those that have superior and desirable qualities are chosen for propagation and introduction as named varieties.

Fig. 34. The acorn shown at *a* is a hybrid of the live oak shown at *b* and the overcup oak shown at *c*. (*Courtesy of Tex. Agr. Expt. Sta.*)

Sports occur quite frequently in sour cherries. The Red Reine plum, from Idaho, and the Washington Navel orange, grown widely in California, originated as bud sports. The Starking variety of apple came into existence as a sport of an apple tree of the Delicious variety. The wide popularity of this variety is made possible by the vegetative propagation of millions of trees, directly or indirectly, from the original limb. The sweetpotato gives rise frequently to off-type sports which are possible sources of new varieties. The same is true for many ornamental plants and flowers. Sports may be found to exist in several different

Fig. 35. Variations in seedling pecans. The Texas Prolific (*a*) and Onliwon (*b*) are seedlings of the San Saba variety (*c*); they can be perpetuated only by asexual means.

Fig. 36. Self-sterile Bruce plum at X, Santa Rosa plum at X, caged with bees at Y to provide for controlled cross-pollination and hybrid seed in Bruce.

stages of growth, and this causes confusion in their identification. Where such a variation shows up on a twig or branch, which may produce fruit of a deeper color or with some other noticeable character, it is not difficult to detect it and propagate it. There would be no positive method of determining this difference, however, except by propagation and field trial. In some rare cases, however, mutant buds have been used in propagation work and a whole tree produced that was different from the type of the variety. Attempts to produce mutations artificially are still in the experimental stage.

Types of Propagation. Briefly, horticultural plants are propagated in either of two principal ways:

Seed. A seed is a reproductive structure that bears heritable factors from a sperm cell and an egg cell. Thus if the pollen which produces the sperm cell that fertilizes the ovule comes from the same variety of a relatively pure line, the seed produced is likely to reproduce true to variety. Many varieties, however, cannot be either self- or close-pollinated because of *self-sterility, dioecism, dichogamy,* or some other inhibiting condition. Cross-pollination occurs freely where plants are pollinated by wind or by insects, and if the pollen comes from a foreign source, the seed in question is not likely to reproduce the variety.

Vegetative Parts. Vegetative propagation differs from seed propagation in that vegetative plant parts, such as stems, roots, bulbs, and leaves, are used rather than seeds. To propagate a new plant from any of these, it is necessary that the vegetative portion either (1) produce new roots that will support it as in cuttage and layerage or (2) unite with another plant that provides the root system, as in budding and grafting.

The methods by which plants may be reproduced vegetatively are bulb propagation, layerage, cuttage, and graftage. These will be considered separately in subsequent chapters and the advantages of each method and the factors that influence a final choice of the one to use for a given plant will be considered.

Briefly, some plants respond readily when propagated by any one of the methods; others can be propagated easily by only one of the methods. For those that can be propagated by more than one, the economy of the method and the usefulness or value of the plants produced are considerations that influence the choice of the method to use.

The general uses and advantages of vegetative propagation can be discussed appropriately under these headings:

PERPETUATION OF VARIETIES. A variety consists of a group of plants with certain fixed qualities or characteristics. Popular varieties are those that have certain combinations of qualities of plant and fruit that make them desirable. The Delicious apple variety, for example, has the desirable tree characteristics of good growth, disease resistance, and regular and heavy production; it has the desirable fruit characteristics of large size, red color, and good taste. Ornamental plants may be valuable because of unusual habits of growth—upright, drooping, or weeping; they also sometimes have an especially desirable characteristic of the leaves or flowers, such an unusual coloration or an attractive pattern of variegation. Such unusual plants will rarely reproduce true to type from seed, for reasons outlined in a previous paragraph. They can, however, be reproduced true to variety by the use of vegetative parts. Parts of the desirable parent plant itself, such as cuttings, buds, grafts, and bulbs, when caused to grow will reproduce the variety. The Elberta peach is an example of a variety that has been perpetuated since 1870 by vegetative methods, principally budding; the millions of trees of the variety that have been propagated and all the peaches that they have borne have been essentially alike. Standard grades and brand of fruits, so essential in efficient, orderly marketing are possible because of vegetative propagation. The marketing of the fruit from a large number of trees of one variety is relatively simple when compared with the problem of marketing the fruit from seedling trees, no two of which produce fruit that is alike. Since fruits from individual seedling trees vary widely in ripening date, size, color, flavor, and other qualities, the important relationship between vegetative propagation and marketing is apparent.

PRODUCTION OF UNIFORM ROOTSTOCK. Vegetative propagation is also important in the production of uniform rootstock for budding and grafting. It is known that the rootstock exerts a direct influence on the top growth of a plant. Seedling plants vary in habits of growth and in fruiting; they vary also in the qualities that determine their value as rootstocks, such as type of root system and resistance to drought, cold, parasites, diseases, and other external influences. The Delicious apple, for example, is practically immune to a serious disease known as collar rot. It can be used to produce a trunk for other apple varieties by graftage, and the resulting tree is immune to collar rot. When seed of the Delicious apple are planted, however, the seedlings that grow from them have no such immunity. Likewise, the Rupestris St. George grape is resistant to grape phylloxera. It is grown from cuttings and used widely as a rootstock wherever grapes are grown. Seedlings of it, on the contrary, are not likely to have the resistance of the parent variety to the insect. The Dog Ridge grape, Myrobalan plum, multiflora rose, and Old Home pear are other examples of plants which are propagated vegetatively and have special merit as rootstocks or body stocks; seedlings of these, on the contrary, are variable and do not have the distinguishing characteristics that make the parent varieties valuable as rootstocks. Some rootstocks produce objectionable suckers freely from seedlings, but do not produce them from rootstocks grown from properly prepared cuttings. In research work with fruit varieties, where comparisons of varieties are being made, or where responses to fertilizers or cultural treatments are being measured, it is particularly important that the rootstocks used be uniform.

Plants That Produce Seedless Fruits. There are many horticultural varieties of fruits and vegetables that do not ordinarily produce viable seeds. Vegetative parts of the plant are used entirely in the propagation of standard varieties of these plants. The banana, Washington Navel orange, Thompson Seedless grape, and the common fig are examples of plants of this class.

Plants That Produce Seeds Difficult to Grow. Some plants can be propagated more economically by vegetative parts than by seed. The seed of the sweet potato and the Irish potato are small. They do not germinate readily unless given special treatment. The seedlings are small, delicate, and susceptible to disease. Even with careful handling they will not make sufficient growth to produce marketable crops in one season. A marketable crop can, however, be grown in one season by using the fleshy root of the sweetpotato to produce "slips" for field planting, and by planting Irish potato tubers, either whole or cut into pieces.

Methods Used to Propagate Horticultural Plants. Most plants can be propagated by more than one method. For the important groups of horticultural plants the methods outlined briefly below are those that are in common use.

Vegetables. Seeds are used principally to propagate named varieties of most of the common vegetables, such as the bean, pea, cabbage, tomato, onion, and others. Some of these crops, beans and peas, for example, are regularly self-pollinated. Tests in California showed that there is very little cross-pollination in bean plants that grow side by side. Seed growers have little difficulty in keeping self-pollinated varieties true to form. Other vegetable crops are regularly cross-pollinated. This is generally true of watermelon, muskmelon, and sweet corn; and it is invariably true of asparagus and spinach, since they are dioecious. Seed growers protect the purity of the seed of these crops by planting and growing plants of a given variety in an isolated location.

Vegetative parts are used to reproduce certain vegetable crops. The Irish potato, sweetpotato, and horseradish are examples of those propagated vegetatively. In addition, vegetative propagation is occasionally used for vegetables that are normally grown from seed. The tomato, for example, is sometimes grown from cuttings to save a promising plant, possibly for further plant breeding; it is occasionally grafted onto other members of the nightshade family to study graft unions and the reciprocal influence of the root system and top. Likewise, cabbage and broccoli, though normally grown from seed, can be readily propagated by leaf-bud cuttings and leaf cuttings. This makes it possible to perpetuate certain outstanding plants for research investigations.

Flowers. The discussion of vegetables above applies in like manner to flowers. Most of the common flowers, the zinnia, phlox, and petunia, for example, are grown from seed. Others, such as the chrysanthemum and canna, are readily propagated by vegetative methods.

Fruits. Some few fruits are propagated by seed. Orchards of tung nut trees are developed by planting the seed. As previously outlined, one of the ways by which new varieties of fruits originate is from *seed.* Seedage is of relatively minor importance, however, for the propagation of most varieties of fruits.

The vegetative methods are used far more widely than seeds in the propagation of fruits. The small fruits are reproduced by suckers, layers, or cuttings. The tree fruits are most generally propagated by budding or grafting onto rootstocks grown from seed or from cuttings.

Ornamental Plants. Some are grown from seed. In general, however, the choice forms are propagated by some form of vegetative propagation. Cuttage, layerage, budding, and grafting are in common use.

QUESTIONS

1. What is a variety?
2. What are the ways by which varieties of horticultural crops originate?
3. Explain why some varieties can be reproduced satisfactorily from seed, while others cannot.
4. What is a sport? What are common ways in which sports differ from the parent plant?
5. How can a variation in a part of a plant be identified as a sport?
6. What are advantages of and reasons for propagating plants from vegetative parts?
7. What important vegetable crops are propagated by seeds? By vegetative parts?
8. Why are fruits, such as apple and peach, generally propagated from vegetative parts?

SUGGESTED REFERENCES

Gardner, V. R.: Studies in the Nature of the Pomological Variety, *Mich. Agr. Expt. Sta. Tech. Bull.* 161, 1938.

Tukey, H. B., and K. D. Brase: Random Notes on Fruit Tree Rootstocks and Plant Propagation, I, *N.Y. (Geneva) Agr. Expt. Sta. Bull.* 649, 1934.

———— and ————: Random Notes on Fruit Tree Rootstocks and Plant Propagation, II, *N.Y. (Geneva) Agr. Expt. Sta. Bull.* 657, 1935.

———— and ————: Random Notes on Fruit Tree Rootstocks and Plant Propagation, III, *N.Y. (Geneva) Agr. Expt. Sta. Bull.* 682, 1938.

Yeager, A. F.: Breeding Improved Horticultural Plants, I, Vegetables, *New Hampshire Agr. Expt. Sta. Bull.* 380, 1950.

————: Breeding Improved Horticultural Plants, II, Fruits, Nuts, and Ornamentals, *New Hampshire Agr. Expt. Sta. Bull.* 383, 1950.

CHAPTER 6

Germination of Seeds

The embryo formed as the result of fertilization normally makes only a limited amount of growth before its development is checked. It becomes dormant as the ovule ripens into a seed and remains in a condition of arrested development within the mature seed until it is subjected to conditions favorable to growth. The process whereby the embryo resumes growth and the radicle and plumule break through the seed coat is known as *germination*.

Water Absorption. For the initiation of germination it is essential that the seed be subjected to a favorable supply of moisture. With most seeds, it is sufficient to place them in a soil or other medium with a suitable moisture content. Some seed, however, which have impervious seed coats, are unable to take up sufficient water for germination without special treatments, discussed later. During germination the embryo, and the endosperm if present, swell and push the seed coat off. This increased water content of the seed is also a necessary condition for the processes of gas interchange, chiefly the intake of oxygen and the giving off of carbon dioxide. Other processes that go on with the increased water content of the seed are the translocation of food and the resumption of activity in the protoplasm.

Movement and Utilization of Foods. The seed contains relatively large quantities of stored food, in such insoluble forms as starch, fat, and protein. In order for these materials to be used by the seed, they must be transformed into soluble compounds. Certain specific enzymes, produced within the cells carry on this process of digestion which renders the stored foods available.

Following digestion, there is a movement of food, from the storage cells to the growing parts of the germinating seed, which are principally the growing points of the radicle and the plumule. The foods are used up in various processes at the growing points, while continued supplies are made available in the storage cells; as a result the movement continues as a process of diffusion.

At the tips of the radicle and the plumule, new cells and tissues are being formed and elongation of others is in progress. The food supply is utilized to form new protoplasm, to thicken the walls of new cells, and to form wood fibers, conducting tissues, and various other specialized products and structures associated with growth.

In the synthesis of organic compounds in the leaf, the kinetic energy derived from the sun was converted into potential energy. Energy of growth is provided in the breaking down of these carbohydrates and other stored foods. This decomposition of stored foods occurs in the process of respiration, which releases water and carbon dioxide, in addition to energy. Only a part of the energy so produced is used in production of new tissues, and a considerable portion is given off as heat.

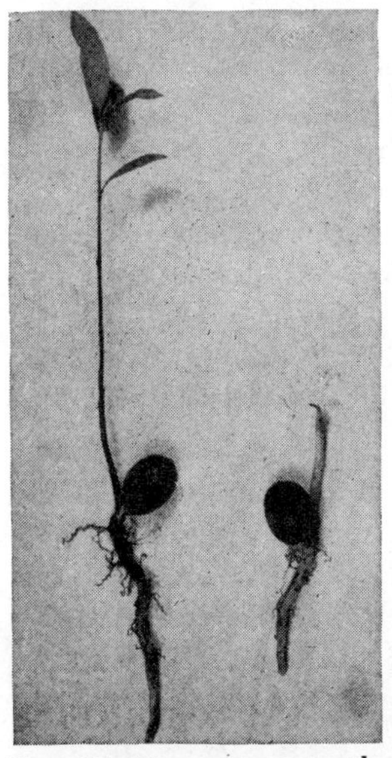

Fig. 37. Two stages in the early growth of seedlings.

Seedlings. The process of germination is considered as complete when the seed coat is broken and the radicle and plumule pass outward from the seed, from which time the young plant is regarded as a *seedling*. The radicle produces the taproot of the young plant, and the *plumule* produces the stem. *Emergence* is the appearance of the young stem above the seedbed surface, and this is of prime interest to a grower. The seedling stage continues until the young plant begins to manufacture its own food and is no longer dependent upon the food stored in the seed. Actually horticultural plants that grow from seed are frequently regarded as *seedlings* for an indefinite period. Thus, tomato, or onion plants, that have made substantial growth over a period of several weeks are called seedlings. Likewise peach, apple, or orange trees grown from seed and allowed to grow for one or many years, and perhaps to bear fruit, are commonly designated as seedlings.

Seedlings of monocotyledonous plants produce a temporary root system that is followed by the permanent system, but in the dicotyledonous seedlings the first root system becomes a part of the permanent one. As the seed of some monocotyledonous plants germinate, the cotyledon is raised to the surface. This happens when the onion seed germinates. In the germination of seed of other monocotyledonous plants, the cotyledon remains in the ground. This is true when corn seed germinates. The same difference occurs in seedlings of the dicotyledonous plants. In melon, radish, bean, and other seedlings of this type, the cotyledons are pushed out of the ground by elongation of the hypocotyl and often function temporarily as leaves. On the other hand, the plumules of seedlings of the garden pea push upward through elongation of the epicotyl, and the cotyledons remain in the ground. It is generally considered that seedlings of this type can emerge from the soil with less difficulty than those which must raise the cotyledons.

Requirements for Germination. In order for seed to germinate they must have proper conditions of moisture, temperature, oxygen, and possibly light for at least some seed, and seed must be viable to respond to these conditions. Anyone of these may become a limiting factor in the germination process.

Moisture. The amount of moisture required for germination is usually that which will completely saturate and soften the seeds. The absorbed water saturates the cell walls and starch grains and fills the living cells of the embryo and all empty spaces that exist in the seed. The amount of water required for saturation varies for different seeds. Corn is saturated by 43 per cent of its weight of water; peas require 107 per cent; and sugar beets with surrounding pericarp and perianth, 120 per cent. This intake of water is accompanied by a great increase in volume of the seed.

Temperature. Temperature has an effect on germination by its influence on (1) the rate of water intake and (2) the speed of the metabolic processes within the seed. There are wide differences in the temperature requirements of seed for germination.

It is common knowledge that seeds fail to respond if the temperature is too low, and it is also true that seeds will not germinate well if the temperature is too high. Growers in the South seeking to grow a fall crop often obtain poor germination and hence poor stands of carrot, lettuce, celery, and other similar crops in late summer and early fall because of the high temperature.

Maximum and minimum temperatures are the highest and lowest, respectively, at which germination will take place. These provide extremes between which germination will occur for given plants. The temperature within the range at which germination will take place within the shortest time is known as the *optimum*.

The rate of germination and emergence is slow at the minimum temperature, and seedlings that are thus produced are likely to be small and stunted. Seeds that germinate and emerge at a temperature near the upper limit produce seedlings that tend to be weak and spindling and highly susceptible to disease. A higher percentage of germination is likely to be obtained at the optimum temperature, and the plants produced are usually strong and sturdy.

Vegetable and Flower Seeds. Seeds of vegetable and flower plants will germinate under a fairly wide range of temperatures.

1. Seeds of the so-called "cool-season" crops will germinate within the approximate temperature range of from 40 to 90°F. The garden pea, spinach, radish, cabbage, and onion are of this group. If the temperature of the soil in which seeds of this group of plants are planted is near either the minimum or maximum range, poor germination will result. A fairly low temperature, about 60°F., is best for celery. Annual Delphinium seed germinate best at about 60°F. and germinate very poorly at 68°F. or above. For the potato 68°F. has been determined to be the best temperature.

2. For the warm-season crops, such as cucumber, muskmelon, and tomato, the temperature range for germination is from 60 to 93°F. or slightly higher. The optimum temperature for germination of seeds of these crops is from 80 to 85°F.

Fruit Trees and Ornamental Plants. Seeds of various fruit trees and ornamental plants will germinate under a wide range of temperature, depending upon the species. Many of them require low temperatures for germination and some will actually germinate near the freezing point. Peach and plum seeds will germinate within the temperature range of 36 to 90°F.; satisfactory germination is obtained where the soil temperature is about 70°F. For apple, pear, and grape, a temperature of 70°F. is satisfactory. Seeds of subtropical and tropical plants germinate best at a slightly higher temperature. Citrus and avocado seeds germinate satisfactorily at about 80°F.

Light. Seeds can be conveniently classified into four groups based upon the influence of light on germination.

1. Light is absolutely essential for the germination of some seeds. This is true for European mistletoe (*Viscum album*) seed.

2. The germination of other seeds is favored or hastened by light, but light is not absolutely necessary for germination. This has been reported to be true for certain varieties of lettuce.

3. Other seeds have their germination inhibited by light. This is true for certain species of wild onions and lilies which germinate best in total darkness.

4. There are also seeds which are indifferent to light and apparently germinate equally well in light or darkness. This is true for seeds of most horticultural crops such as cabbage, beans, peas, pecans, peach, and blackberry.

Oxygen. The seed is a living structure and requires oxygen. The embryo requires oxygen for respiration and the initiation of growth. Under ordinary conditions the seed does not suffer for lack of this element, since the atmosphere contains an abundance. In compact, poorly prepared, or excessively wet seedbeds, germination may be retarded or prevented by a lack of oxygen. It has been shown conclusively with asparagus that water in excess is not detrimental, but that oxygen is required for germination. Seeds in unaerated water showed no germination at all in 2 months, though others, in water through which air bubbled, germinated in 1 week.

Viability and Vitality. A seed is viable if it is capable of germinating. Vitality refers to the vigor or strength possessed by the seed for growth. Viable seeds vary in vitality, being influenced largely by these factors:

Vigor of Parent Plant. Seed from weak plants are apt to be deficient in stored foods and also to have small embryos. They will produce less vigorous seedlings than those from normal plants. Atmospheric humidity and temperature during seed development may also influence vitality. Dry atmosphere and absence of very low temperatures are favorable conditions. It is important that seed for planting mature normally on the parent plant. Seeds that are harvested before they mature may germinate, but they are less likely to do so under unfavorable conditions. They also lose their viability more quickly than do seeds that are fully mature. It is necessary to harvest some seeds before they become dry, to prevent shattering; these should be allowed to become as mature as is consistent with good handling practice.

Age of Seed. The vitality of seeds of the different species is influenced in varying degrees by age. Seeds of the willow will lose their viability in a few days, and those of many tropical plants remain viable only a short time. Among the common vegetable and flower seeds, there is a great variation in this respect. Cucumber, endive, celery, and chicory will often germinate satisfactorily for 8 to 10 years, but dandelion, martynia, onion, parsnip, and sweet corn will ordinarily last only 1 to 2 years, as commonly stored.

Storage of Seed. Different kinds of seed remain viable longest and ultimately germinate best if stored under conditions especially suited to their specific requirements. This should be recognized in providing suitable storage. Briefly the principal variable factors of storage are humidity, temperature, and oxygen.

Respiration is one of the principal causes of deterioration. It is the process by which carbohydrates and other organic materials in the seed are reduced to carbon dioxide and water. Respiration occurs most rapidly under conditions of high temperature and high humidity.

In semitropical and tropical countries difficulty is often encountered in preserving seeds from one season until the next because of the prevailing high temperature and high humidity. The difficulty can be partially overcome by storing the seed under dry conditions in sealed containers to prevent the absorption of moisture. In Texas, onion seed stored in cloth bags at room temperature remained viable for only about 20 months, while those that were sealed in glass jars either at normal pressure or in vacuum remained viable for about 4 years. Onion seed in the same test that were sealed in glass jars under vacuum and kept at about 36°F. remained viable for 11 years.

The two ways to retard respiration are to maintain a low moisture content and a low temperature. A combination of coolness and of dryness thus can be used effectively to store most vegetable and flower seeds. Between the two factors, dryness is probably more important than coolness.

FIG. 38. The onion seedlings in the flat on the right grew from seed which had been kept sealed in air-tight containers for 3 years. Onion seed stored during the same period in cloth bags failed to germinate when planted in the flat on the left.

Some seeds, however, must be stored moist to preserve their viability. This is true of apple and pear seed. It then becomes necessary to store them at a low temperature to retard respiration and, incidentally, to prevent mold and decay. Certain seeds, such as the acorn, which must be kept moist, will sprout prematurely in storage, unless a low storage temperature is maintained to prevent it. Storage temperatures may vary considerably, but those in the 30 to 40°F. range are effective in preserving the viability of most seeds if other environmental factors are likewise satisfactory. The seed is a living organism and as such requires oxygen. Packing seed in moss or other material that is too wet will result in inadequate aeration and the consequent death of the seed.

Seeds of most *fleshy fruits* remain viable longest if they are kept *moist* and *cool* during storage. The peach, apricot, plum, apple, and pear are examples of seeds which require these conditions. They may be handled in dry condition for convenience and for economy in transportation. But from shortly after harvest until planting they are commonly stored so that they will be continually moist and cold. For large seeds, such as peach, plum, and apricot, the moist condition is provided by packing the seed in moist peat moss or similar material or by stratifying them in soil outside. The cool temperature is provided by placing the packed or stratified seed outside in regions where the winters are naturally cold, or by placing the packed seed in cold storage where the winters are mild. Apple and pear seed for planting should be stored under cool, moist conditions as soon as extracted from the fruit. It is a common practice to place them in loosely filled cotton bags and store between cakes of ice. Citrus seeds deteriorate rapidly if allowed to become dry. They then must be kept moist, and under this condition they will germinate unless the temperature is kept low. Hence, for prolonged storage, they are kept at about 32°F.

Seeds of certain *dry fruits* likewise lose their viability if allowed to become and remain dry for a long period. The acorn and chestnut are examples of seeds of this class. It is necessary to keep them continuously moist. Since they will germinate at warm temperatures, it is necessary that the storage temperature be maintained at near 32°F. to keep them dormant for a prolonged period.

Seed of *nuts*, as the walnut and pecan, contain a high percentage of oil. They become rancid in time at ordinary temperatures and the seed lose their viability when this happens. Rancidity can be prevented and the seed will remain viable for 2 or more years if stored at a temperature of about 32°F.

Seeds of ornamental trees and shrubs are variable in their storage requirements. Some, such as the holly and magnolia, require moist storage. Others, such as the ligustrums and mimosa, will remain viable if stored dry. With both, a low temperature is best.

It is clear from the above discussion that all seeds require oxygen during storage; that some must be stored moist and others dry; and that low temperature is helpful in the storing of seed because it restricts respiration, it retards germination of those seeds that must be stored moist, and it retards rancidity.

Seeds that must be stored moist should be packed in moss that is loose, fibrous, and well aerated.

Delayed Germination: Causes and Treatments. In the preceding discussion it is presumed that seeds will germinate if viable and if subjected to favorable environmental conditions. Most seeds will do that. Avocado, pecan, orange, and grapefruit seeds, for example, frequently germinate in the fruit on the tree before harvest. Watermelon and eggplant seed will germinate promptly as soon as the fruits that produce them are mature. With the viable seeds of other plants there is a long delay in germination, and oftentimes ultimate failure, even under favorable conditions. The delay can be shortened by certain treatments that enable the seeds to germinate. Some of the factors and conditions that tend to delay germination, and treatments that overcome them, are discussed in the following paragraphs.

The Seed Covering. The true seed is the mature ovule; its immediate covering is the testa. In some seeds it is thin and delicate. In others it may be so tough and impervious as to interfere with germination. Such is the case in seeds of the mimosa and persimmon. The endosperm of some seeds encloses the embryo so completely that germination cannot proceed normally. Such is the case in seeds of the date and redbud. The seed may be enclosed in other structures which interfere with germination. The shells of the pecan and of the hickory, for example, develop from the ovary wall and are hence the pericarp. The bony seed covering of the peach, almond, and blackberry develops from the endocarp of the ovary wall. Beet seeds are enclosed in the coalesced pericarps and calyx bracts of the inflorescence. The seed coat, alone or in combination with the various other structures that enclose the embryo, may delay germination in any one of several different ways.

Moisture Absorption. The moisture necessary for germination is fully effective only when it is absorbed by the embryo. It is not sufficient that moisture surrounds the seed coat. Seeds of alfalfa, olive, walnut, mimosa, and tung have coverings which are more or less impervious to moisture. Hence they restrict the rate of water absorption and delay germination in doing so.

Oxygen Intake and Products of Metabolism. In some cases tissues that surround the seed may delay germination by retarding the passage of oxygen into the region of the embryo. This accounts for the poor germination of freshly harvested lettuce seed of certain varieties, particularly at a temperature as high as 86°F. Spinach seed does not germinate well at high temperatures. It has been suggested that products of metabolism that arise and accumulate only at higher temperatures are responsible for this failure to germinate.

Strong Coverings. Some very strong seed coverings are known actually to prevent emergence of the radicle and plumule of the embryo, even though other conditions are favorable for germination. This is known to occur in the peach and walnut. Tests have shown that black walnuts required an internal pressure of over 627 pounds per square inch to break the shells.

Toxic Materials. It has been shown that the covering of certain seeds contains toxic materials which inhibit germination. This is true in certain varieties of sugar beet.

Ways to Weaken Seed Coverings. Some of the treatments used for seeds with strong or impervious coverings are outlined below.

Mechanical Means. Hard or strong seed coats, regardless of their structure, can be altered to permit germination by certain mechanical treatments. Peach pits are regularly cracked with a hammer before planting, if necessary to weaken the hard endocarp, without injury to the embryo. Mimosa seed will germinate more readily if the hard seed coat is nicked with a knife or scratched with an abrasive. Clipping the end opposite the micropyle permits ready absorption of water and quick germination of some seeds. This is used effectively on olive seeds, which germinate poorly unless subjected to some treatment that will render the seed more permeable to water. The hard coverings of seeds can be weakened by revolving them in a container with an abrasive.

Chemical Methods. Chemicals are also used to break down the seed covering and render it more pervious to water. Sulfuric acid, used in strengths varying from concentrated to dilutions of 50 and 25 per cent, has been found effective. The seeds are submerged in a solution of the chemical for a specified time, which is determined by the strength of the solution and the toughness or resistance of the seed coat. Potassium hydroxide and hydrochloric acid are also used for some seeds. In every case it is a safe precaution to wash the excess solution from the seed after treatment. Examples of seed treatments which hasten germination by weakening the hard seed covering are outlined below:

1. Untreated seeds of sweetpotato germinate slowly when planted, or fail entirely. Good and prompt germination can be obtained by soaking the seeds before planting in concentrated sulfuric acid for 20 minutes.

2. The strawberry seed when mature is enclosed in a dry impervious pericarp. It germinates slowly if planted without treatment. Soaking in concentrated sulfuric acid for 15 minutes before planting hastens germination.

3. Blackberry seeds are slow to germinate when planted without special treatment. This is considered to be due to the high breaking strength of the endocarp which encloses the seed. Germination can be hastened by treating the seeds for 1 hour in concentrated sulfuric acid, followed by moist storage at a cool temperature.

4. Spinach seed germinate slowly, and in the meantime an unfavorable condition such as drying may develop in the seed-bed which will prevent ultimate germination. Soaking the seeds in a solution of equal parts of concentrated sulfuric acid and water for 30 minutes will hasten germination and in many cases produce a better stand.

5. Germination of redbud seeds is hastened by treatment in concentrated sulfuric acid for 20 minutes followed by stratification for 50 to 60 days.

6. Treating American holly seed for 5 minutes with a normal solution of potassium hydroxide and then for 3 minutes with a normal solution of hydrochloric acid will increase permeability and thereby hasten germination.

Soaking. Soaking seeds in water for varying periods of time up to 4 or 5 days will, in many cases, hasten germination. This would be expected, since water intake is the first step in the process. For prolonged soaking, some means of providing aeration should be used. This can be done by changing the water daily, or by having seed in running water. Warm water is absorbed more rapidly. Asparagus seeds soaked in water at

86°F. took up their maximum amount of water (about 43 per cent) in 35 hours, while those in water at 64°F. required 65 hours to absorb the same amount.

The following soaking treatments are commonly used for the seeds specified:

Kind of seed	*Soaking period*
Celery	1 day
Apple	2 days
Asparagus	3 to 5 days
Mulberry	4 days
Pecan	4 days
Osage orange	5 to 7 days

Some seeds are treated with hot water to facilitate the absorption of moisture by the seed. A common procedure is to heat the water to 180 to 200°F. and plunge the seed into it. Water and seeds are allowed to cool together. Boiling water (212°F.) is also used. The seed, contained in a cloth bag, are plunged in the boiling water and left there for a period up to 2 minutes, the actual time varying with the seed.

Stratification. Another treatment that influences germination by its effects on the seed coat is stratification. The actual method of stratifying seeds may be varied in several ways. The seed may be placed in layers, alternating with layers of sand, in a large box that is left open at the top for the addition of more water as needed, or they may be placed in a shallow pit or trench and covered with earth. The one precaution necessary is to place them where there will be sufficient drainage to keep the soil from becoming waterlogged. The soil or sand should be moist but not saturated if the seed is to remain in it for an extended period of time. Stratification keeps the seed covering moist, which permits readier absorption of moisture when the seeds are planted. Stratification is one way of preserving seeds that lose their viability if allowed to remain dry for a considerable time. Seeds of apple, pear, and cherry belong in this class.

Moisture is necessary for the proper afterripening of certain seeds, discussed in a later paragraph, and stratification is a convenient way of providing this moist condition.

Stratification is used more widely than any other preplanting treatment to preserve viability and hasten germination. Seeds of the following plants and many others are commonly stratified for a period prior to planting: peach, plum, cherry, apple, pear, quince, grape, persimmon, hickory, blackberry, strawberry, rose, pine, magnolia, and oak.

Afterripening Processes. Seeds of many species of plants will not germinate until they have undergone changes that, for convenience, are known as *afterripening processes.* Afterripening is defined as the changes which take place in the seed, which make it ready to sprout. In general these changes occur in seeds most readily if they are in a moist medium and kept within a temperature range of about 33 to 40°F. Failure to provide these conditions results in *delayed germination* when the seeds are planted.

In practice, it is customary to stratify the seed in moist sand, peat moss, or some similar material, and place them where the temperature through a definite period of time will be within the range effective for afterripening.

Delayed germination often prevents the germination of seeds when they would likely be killed by winter freezes. It also prevents the loss of fruits and seeds by preharvest sprouting of the seed.

Two different types of afterripening are recognized:

1. *The Rest Period.* The time during which mature seeds will not germinate, under favorable conditions, but will retain their viability and perhaps germinate at some future date, is known as the *rest period.* The *dormant period* of a seed, on the contrary, is the period between maturity and the time when the seed finally germinates. The rest period of most seeds normally lasts only a few weeks; the dormant period may continue for years. The most effective treatment for breaking the rest period of most seeds is to stratify seed in a moist medium at a cool temperature, as outlined in a previous paragraph (see page 78). Seeds of various species properly stratified require different periods to complete the rest period. If they are not kept moist and cool, they will not be capable of germinating for an extended period. Below are listed representative plants, the seeds of which have a rest period and the approximate length of time required for the ending of the rest period when kept moist and cool.

Kind of plant	*Length of rest period, days*
Pear	55 to 65
Apple	60 to 75
Plum	60 to 90
Peach	75 to 100
Grape	90 to 140
Blackberry	120 to 150
Magnolia	120 to 150

Not all seeds, however, are characterized so definitely by *rest-period* phenomena. These are examples of seed which have no rest period, or a very short one, and which will germinate whenever conditions are favorable: orange, grapefruit, pecan, mulberry, watermelon, tomato, and eggplant. Many others could be cited.

2. *Immature Embryos.* At the time the seeds and the fruits that bear them are apparently ripe, the embryos of seeds of certain species of plants are not mature. This is the reason for delayed germination of carrot and of several species of holly. Apparently little can be done to hasten the maturity of such embryos. It is important to hold them under conditions of temperature and humidity that will help preserve their viability until such a time that the embryos complete their growth and will germinate. Carrot seed will normally complete their development in about 90 days. Holly seeds require from 18 months to 3 years, and in the meantime, they should be stratified in a moist, cool place.

Seed Treatments to Control Disease. In the following discussion, the term *seed* is used to designate a matured ovule, a structure that contains an embryonic plant. In some cases they may be *true seeds,* in others they may be *fruits* in which the seeds are included, as, for example, those of the beet or lettuce. Tubers, fleshy roots, bulbs, and other vegetative structures that may be used for reproduction and are, hence, sometimes referred to as "seed" will be considered in another chapter.

Many diseases are transmitted by the seed, the organism being borne either upon the surface or within the seed coat. Seed-borne diseases may attack the very young seedling plants either before or shortly after they emerge, and prevent a good stand;

in other cases the damage may be delayed until the plant approaches maturity.

Diseases Borne on the Seed Coat. Organisms that produce black rot of cabbage, smut of sweet corn, anthracnose of watermelon or muskmelon, and the fungi that cause damping-off of many kinds of seedlings are examples of diseases that may be borne on the seed coat. There are many others that might be mentioned. The seeds may become infected in several ways; probably the most common source is diseased mother plants. Spores of the fungi cling to the seed coat or are embedded in it. When the seed germinates, the fungus spore also germinates. In the case of damping-off the disease shows up at once. In others—anthracnose of watermelon, for example—the presence of the disease may not be apparent until the plants have made considerable growth. These diseases and others may be disseminated in ways other than on the seed. Seed treatments for them, however, are considered to be worthwhile precautionary measures. Suggestions for the use of some of the more common disinfectant and protectants are given in the following discussion:

Bichloride of Mercury. This chemical is considered to be one of the most effective disinfectants for seed. The solution usually recommended is made by adding 1 gram of the bichloride to 1,000 cubic centimeters of water. A solution of the same strength may be prepared in greater volume by dissolving 1 ounce of the bichloride crystals in 8 gallons of water. Soaking the seed for 15 minutes is recommended to control diseases of cabbage and related plants; 10 minutes for diseases of cantaloupe, watermelon, squash, and cucumbers; and 8 minutes for tomato, pepper, and eggplant. After the treatment is completed, the seed should be rinsed several times in water. Metal containers should not be used for the treatment; glass, stoneware, or wooden vessels may be used. This chemical is extremely poisonous and should be used with caution.

Copper Sulfate. Soaking seed in a solution of this chemical is effective in destroying certain seed-borne diseases. It has been recommended particularly for destroying the damping-off organism on spinach and tomato seed. The solution is prepared by adding 1 to 2 ounces of the copper sulfate to 1 gallon of water.

The seed are soaked for approximately 1 hour, after which they are dried without rinsing preparatory to planting.

Formaldehyde. This chemical has long been used in liquid form, properly diluted, as a seed disinfectant for the smuts of cereals. The solution is prepared by adding 40 per cent formalin (formaldehyde) to water at the rate of 1 pint to 30 to 40 gallons of water. Seeds to be treated are placed in a cloth bag and immersed in the solution. Most vegetable seeds are treated for 10 minutes, after which they are rinsed in water or a milk-of-lime solution prepared by adding 1 pound of quicklime to 10 gallons of water.

Other Chemicals. Red oxide of copper in powder or dust form has been shown to be an effective fungicide for damping-off. It has sticking qualities that enable it to adhere remarkably well to the seed coat. Small quantities of seed may be treated by shaking the seed in a closed container to which red oxide of copper has been added at the rate of 1 teaspoonful of dust to each pound of seed to be treated.

Copper carbonate may be used effectively for the control of the damping-off of vegetable and flower seedlings. It is also used as a treatment for wheat smut. It is used as a dust, 2 to 3 ounces being sufficient for a bushel of seed. Larger amounts are not injurious. Seed may be treated several months before planting.

Various commercial preparations are used effectively in treating diseases borne on the seed coat. Examples are those produced under the trade names of Semesan, Ceresan, and Spergon.

Diseases Borne within the Seed Coat. Certain plant diseases are caused by organisms that may be borne within the seed coat. The fungus that causes blackleg of cabbage and related plants is of this class. Treatment of infected seed with chemicals is effective only to the extent of killing the parasite on the outside. Treatment with hot water at a temperature of 122°F. for 25 to 30 minutes is effective in destroying the parasite on the inside as well as on the outside. The treatment is severe, and only seeds with strong vitality withstand it. If seeds are known to be infected, they should be discarded. The treatment is recommended only as a precautionary measure.

The fungus that causes anthracnose of beans is also borne within the seed coat; hence seed treatments used to control surface-borne parasites are not fully effective. The bean seed is killed by hot-water treatment of intensity and duration that kill the fungus, and means of control other than seed treatments must be relied upon.

Virus Disease. In addition to the diseases caused by visible parasites, on or within the seed coat, there are also virus diseases. Little is known about the real nature of these, and no good classification on them has been made. They may be transmitted from one plant to another in different ways. Most viruses are not carried by the seed from diseased plants. Certain mosaics of legumes, lettuce, wild cucumber, and other plants are, however, transmitted by the seed from infected parent plants.

The cause of virus diseases is not known; hence no treatments are recommended that will make infected seeds safe for planting.

Seed Growing. Since the discoveries of sex in plants by Camerarius and of the fundamental laws of inheritance by Mendel, it has been recognized that considerable knowledge and skill are required for the intelligent production of seed. The inherent possibilities of mature plants are contained in the seeds from which they grow, and from this fact comes the old saying that nothing is so costly as cheap seed.

Methods of Production. The commercial production of vegetable seed involves two distinct steps: first, the production and standardization of stock seed; and, second, the growing of these seed on a large scale for the trade.

The stock seed are usually grown by the seed firms themselves, on their own grounds. Methods employed are determined, to some extent, by the characteristics of the plant under consideration. The same procedure would not be followed, for example, on the bean and the muskmelon. The breeding plots are kept under careful observation; all weak, diseased, or off-type plants are removed whenever noted; and special emphasis is placed on vigor, season, quality, uniformity, and trueness to type.

Stock seeds produced by the seedsman are then sent out to contract growers. These are generally farmers of a part of the country where seeds of the crop in question can be grown. Different sections are adapted to the growing of different kinds of seeds, and these may be remote from the centers of commercial production of the same vegetables or flowers. The fieldmen for the seed firms visit the farms of the contract growers regularly and make inspections of the crops. All offtype plants are removed, a procedure known as *roguing*, and the purity and trueness of type of each variety is observed. In species that hybridize readily, each variety must be planted far enough from all other varieties to ensure that it will not be crossed. The crop of seed is then harvested and sent to the seed dealer, who offers it for sale

Fig. 39. Two common ways of making a germination test—sand flat on left, rag-doll method on right.

the following season and plants a plot on his trial grounds at the same time, as a check.

Classes of Seeds. The botanical purity of some seeds can be determined by an examination of the specimens themselves; others can be judged only by the plants that they produce. Most vegetable and flower seeds fall within this latter class. The most careful examination reveals no consistent differences between the seeds of cauliflower and cabbage, for example. Marked resemblance exists between the seeds of carrot and parsley, cucumber and cantaloupe, onion and leek, and pepper and eggplant. Growers cannot determine the quality of a sample of seed by inspection. They must depend partly on the reliability of the seedsman from whom they are purchased, and they may, in addition, run seed tests to determine the value of a given sample.

Seed Testing. The testing of seed involves two separate considerations:

Mechanical Analysis. A sample of seed may show a considerable amount of inert material or even dead seeds of some other kind. It may contain viable seeds of other plants, including noxious weeds. The Russian thistle was first introduced into the

Fig. 40. Field of onions being grown for the production of seed.

United States mixed with a shipment of wheat seed; it has become one of the most dreaded plant pests. The purity of any sample of seeds from this viewpoint must be determined, especially when seeds are bought in bulk and in large quantities.

Physiological Examination. There are three phases to this step in the testing of seeds: (1) Germination tests are conducted, in which various methods may be employed. Moist cloth, saucers, and blotting paper may be used, but it is preferable to plant the seeds under the more normal condition of a sand or soil plot. Regardless of the method, it is important to ascertain the percentage of seeds that germinate and the time required. Such information will determine whether seeds are to be planted normally, planted thicker than usual, or discarded. (2) Another phase of the physiological examination involves testing the vitality of the sample. Vigor of growth of the young seedlings is a very important character. The young plant is dependent for a time on food stored in the seed, and for this reason healthy mature seed will give the plant a better start. Seeds may be sown in plots of clean sand and tested for vitality at the same time the germination test is being made. The presence of disease in the seed may be detected if the seeds are sown in sterile sand. Diseased seeds are frequently responsible for losses that could have been avoided by proper testing. Seeds with pronounced vitality will frequently yield a good stand under unfavorable conditions, whereas those that do not have such unusual vigor would scarcely germinate under the same conditions. (3) Additional information on the value of a given lot of seed for planting may be gained by growing a trial crop from a sample of the seed. This is done one season, and if the resulting crop is satisfactory, seeds from the lot are used for planting during seasons that follow. The growing of such test crops may yield information relative to the presence or absence of seed-borne diseases, trueness to strain or variety, and the presence or absence of closely related species or noxious plants.

QUESTIONS

1. What are the different interpretations of the term seedling?
2. What is the difference between germination and emergence?
3. What is the explanation for the phenomenon whereby cotyledons of the bean are borne to the surface during germination, while those of the garden pea remain below ground?
4. What conditions are essential for germination?
5. What is the temperature range for the germination of specified seeds?
6. What conditions of storage are best for vegetable seeds? For seeds of fleshy fruits? For oily seeds? For citrus?
7. What are the causes of delayed germination of seeds? What can be done to overcome each cause?
8. What treatments are used for diseases borne on the seed coat? Within the seed coat?
9. Outline ways of testing seeds to determine their quality.

SUGGESTED REFERENCES

Afanasiev, M.: Propagation of Trees and Shrubs by Seed, *Okla. Agr. Expt. Sta. Cir.* 126, 1942.

Bakke, A. L., H. W. Richey, and Kenneth Reeves: Germination and Storage of Apple Seeds, *Iowa Agr. Expt. Sta. Res. Bull.* 97, 1926.

Borthwick, H. A.: Factors Influencing the Rate of Germination of the Seed of *Asparagus officinalis, Calif. Agr. Expt. Sta. Tech. Paper* 18, 1925.

———— and W. W. Robbins: Lettuce Seed and Its Germination, *Hilgardia,* **3:**275–304, 1928.

Chadwick, L. C.: Improved Practices in Propagation by Seed, Herbst Brothers Booklet, 92 Warren St., New York.

Franklin, DeLance F.: Growing Carrot Seed in Idaho, *Idaho Agr. Expt. Sta. Bull.* 294, 1953.

Gross, W. L.: The Vitality of Buried Seeds, *J. Agr. Research,* **29:**362, 1924.

Haskell, R. J.: Vegetable Seed Treatments, *U.S. Dept. Agr. Farmers' Bull.* 1862, 1940.

McLaughlin, J. Harvey: Vegetable Seed Treatments for Oklahoma, *Okla. Agr. Expt. Sta. Bull.* 293, 1946.

Schudel, H. L.: Vegetable Seed Production in Oregon, *Oregon Agr. Expt. Sta. Bull.* 512, 1952.

Tisdale, W. B., A. N. Brooks, and G. R. Townsend: Dust Treatments for Vegetable Seed, *Florida Agr. Expt. Sta. Bull.* 412, 1945.

Toole, Eben H., Vivian Kearns Toole, and E. H. German: Vegetable-seed Storage as Affected by Temperature and Relative Humidity, *U.S. Dept. Agr. Tech. Bull.* 972, 1948.

CHAPTER 7

Methods of Seedage

Horticultural crops are grown from seed for three principal uses. These are to produce commercial crops, to develop new varieties, and to grow rootstocks for budding and grafting.

Seedage Methods for Vegetable Crops. Some vegetable crops are started by planting the seed in the field where the plants grow to maturity. Others are started in special seedbeds and the seedlings are transplanted to the field when they have reached a proper stage of development. Economic factors, botanical characteristics, and necessary cultural treatments determine the system to be used with a given vegetable crop.

Field Seeding. Carrots, cucumbers, beets, radishes, beans, and spinach are examples of a few vegetables that are regularly started by planting seed directly into the field; many others could be cited. A relatively large number of plants are required for a given area. The labor and expense of transplanting are great. It is thus not economically profitable to transplant these crops because the yield and hence the financial returns from an individual plant are low.

Some vegetable crops respond poorly and either die or become reestablished very slowly, when transplanted by ordinary methods. This is true with beans, sweet corn, cucumbers, and watermelons; and these are examples of crops which by prevailing practice are planted directly in the field because of some physiological or anatomical characteristic.

Successful production of any crop depends upon a good stand. It is appropriate, then, to consider some of the conditions that affect the germination and early growth of vegetables that are planted directly in the field.

92

1. DEPTH OF PLANTING. The seed of many vegetable crops are small. The number of seed per ounce for turnips is 10,000, and for carrots it is 20,000. If such seeds are planted deep, the energy of the young plant will have become largely depleted before the plumule emerges. On the contrary, if they are planted too shallow, there is the possibility and likelihood that the seedbed will dry out and leave the seed with a scant supply of moisture before it germinates. Whether the seed are planted deep or shallow, heavy rainstorms may result in packing the soil. Upon drying, a crust forms which interferes with aeration and tends to cause a low supply of oxygen, which is essential for germination. The crust may, in addition, be so strong that it restricts the physical growth of the germinating seed. These various possibilities suggest that problems are involved in planting seeds, particularly small seed, in a way that will reasonably ensure a good stand. The problem is less serious if the soil has good water-holding capacity, if it is loose and friable and does not pack readily. It is also less serious if the seed germinates by elongation of the epicotyl, whereby the cotyledon or cotyledons remain in the soil; or if the seeds are large and consequently have the vitality to germinate and grow under adverse conditions. Obviously deeper planting is more permissible on light friable soils than on heavy compact ones.

2. SOIL-BORNE DISEASES. The fungi which cause damping-off and death of young plants are frequently present and sufficiently active under field conditions to cause poor stands. Thick seeding is not effective in preventing the loss of stand, since damping-off is more serious on plants that grow close together, because of the shading effect and limited ventilation. The disease becomes transmitted from one plant to another more readily if the plants are crowded. Soil-sterilization units for use under field conditions are available and useful to combat the fungi that cause damping-off. In one such sterilizer a kerosene flame is used to heat the soil and reduce the prevalence of the destructive organisms in a narrow band where the seed are planted.

3. SEEDING TO PRODUCE A DESIRED UNIFORM STAND. The field spacing of plants must be such as to produce maximum yields of a high-quality product. If too close, the quality is impaired; and if too great, yields will be low. Precision planters now available enable growers to plant with a reasonable expectation of obtain-

ing a proper spacing. The proper spacing of very small seeds presents a problem. The difficulty with these can be largely overcome by "pelleting." This is a treatment whereby each seed is enclosed in a clay compound. The pelleted seed has such uniform size that it can be spaced accurately when planted by machine. Incidentally, preliminary results show that there is a slight delay in the germination of pelleted seed. This conceivably could be due to the somewhat slower rate of water or oxygen absorption through the coating of clay.

FIG. 41. Two-unit precision planter provides for multiple rows on a bed and uniform spacing for small seeds.

Seed of the beet and related plants present a special problem in obtaining a properly spaced stand. The so-called beet "seed" is really one fruit or several with adhering perianths, containing many seeds. These seed-bearing structures are frequently broken mechanically into pieces before they are planted, so that each broken portion contains a smaller number of seeds. Still, some are likely to contain more than one seed, and when planted, several seedlings will develop in spots. Beets of the best market quality for size, shape, and smoothness will be produced if individual plants have a minimum spacing of about 4 inches in the row. To obtain such a uniform spacing, hand thinning is usually necessary.

4. INFLUENCE OF UNFAVORABLE TEMPERATURE. In some cases, soil and atmospheric temperature become limiting factors in obtaining a good stand of vegetable plants. Lettuce, for example,

is commonly field-seeded to produce the commercial crop. In the Southwest, to have it ready for harvest at the right season, it is usually necessary to plant it when the soil and atmospheric temperatures are too high for best germination. Good germination and, hence, a good stand can be encouraged by holding the seeds in a moist condition at a favorable temperature for 3 or 4 days prior to planting.

Planting in Special Beds. Seedbeds of various sizes are used for starting vegetable plants. When the young seedlings have reached an appropriate size, they are then transplanted to their permanent location, either nearby or in a distant area. There are several advantages to be derived from starting plants in a special seedbed:

1. The grower is enabled to produce a marketable crop of a frost-tender species, such as the tomato, much earlier than if the seed were planted directly in the field. Young seedlings 1 month or 6 weeks old are available for planting at a time when otherwise it would be necessary to plant seed. In any case, whether the crop is for home use or for market, the matter of earliness of maturity is most important. Tomato plants, for example, are grown extensively in south Georgia. Approximately 300,000 plants can be grown on an acre and the annual acreage of seedbeds ranges from 8,000 to 10,000 acres. Most of the plants are grown in 18-inch rows, and a spacing of 10 to 16 plants per foot is common. Stocky plants about 8 inches tall are preferred by the commercial growers in the North who buy the plants. When they have reached marketable size, they are shipped by refrigerator car or truck to the areas where they are to be planted.

2. The practice makes it possible to lengthen the favorable growing season of a locality and thus give a tender species time to mature where the season might otherwise be too short. The eggplant, for example, requires a long, hot growing season; it cannot be grown successfully in some sections of the North unless its season is lengthened artificially by starting the plants in a hotbed or greenhouse. Cabbage, on the contrary, requires a cool growing season. In the South, the weather often becomes too warm as it approaches maturity for the development of highest quality. This can be offset by starting the plants earlier and thus causing the cabbage to mature at an earlier date, when cool weather is expected. For the fall crop of cabbage in the South,

the crop must be started early in order to mature before the winter freezes. Since the weather is too hot for optimum growth of the tender young plants in the open at that time, the slatted frame may be used to a good advantage.

3. The crop occupies the land for a shorter period of time and the grower has opportunity to make more effective use of a certain area. Two or more crops may be grown on the same land within a given time instead of one crop. A green-manure crop may be given time to make better growth, or possibly to decompose more completely, if the plants that are to follow do not have to go into the field so early.

4. Young plants can be given better care in a specially prepared seedbed; the weak ones may be culled out, and the resulting stand in the field will be more uniform both in size of plants and in spacing. Watering and insect control are more easily accomplished in a seedbed than in the open field. Conditions in a seedbed encourage the growth of good plants, since the detrimental effects of drought, excessive rainfall, or insect damage may be largely overcome.

The commercial onion crop is usually grown by planting either dry sets, or green seedlings in the field. In either case the plants are grown in a special seedbed. For green seedlings, in the South, seeds are commonly planted during early September and the young plants will be ready to transplant to the field in November. From 15 to 20 pounds of seed are planted per acre of seedbed. For dry sets the seeds are planted at the rate of about 70 pounds per acre in a seedbed and the young plants are allowed to grow one full season. Crowding of the plants causes the bulbs to mature when they are from $\frac{1}{2}$ to 1 inch in diameter. They are harvested, stored, and planted at the appropriate time to grow a commercial crop of onions the following year.

Celery crops are invariably started by planting the seeds in a special seedbed. When the young seedlings are from 8 to 12 weeks old, they are then transplanted to a permanent location in the field where they produce the commercial crop.

These are the reasons for the prevailing practice of starting celery in seedbeds: celery seeds are extremely small, they must be planted with care to obtain good germination, and they are expensive. In addition to these, it is important to have a perfect stand in the field, to facilitate blanching, particularly with certain varieties. This can be obtained fairly easily by transplanting, but field seeding is uncertain.

5. The saving effected in seed may be considerable. Some vegetable and flower seeds are comparatively expensive, and it is desirable to make economical use of them. When seeds are planted in beds, every strong plant may be used; while if seeding is done directly in the field, the seed must be planted thickly and the seedlings thinned if a good stand is to be obtained.

Fig. 42. The proper way to prepare and plant a seedbed in a flat.

Seedbeds of varying sizes are used for starting vegetable plants. In greenhouse operations, flats are in common use. Likewise cold frames, hotbeds, and outside field beds are used. With all these, successful production of uniform, stocky plants depends upon attention to certain necessary procedures. Seedbeds, regardless of size, should be prepared well and made level. This will facilitate uniform watering of the seedbed and of the young plants. The depth of planting should be uniform. This can be ensured to a reasonable degree by the use of row markers which make a V-shaped furrow. A uniform depth of planting encourages uniform emergence and growth. For tomato, cabbage, onion, and similar crops a planting depth of $\frac{1}{2}$ inch is common. The common rate of seeding in the row is 6 or 7 seeds per inch.

In many cases the young plants are moved directly from the plant bed to the field. Instead of this, the young plants may be transplanted first to pots, paper squares, or similar containers, then, after a period of growth to the field. They may be transferred to flats or cold frames, with more space; and after a further period of growth, to the field. In some cases, seeds are planted directly into containers, and after a period of growth the plants are shifted to the field. This practice eliminates one transplanting operation. It also is a practical means of starting

Fig. 43. Cold frames used for hardening cabbage plants prior to field planting.

crops like muskmelons and watermelons which cannot be moved satisfactorily if the roots are greatly disturbed, but which can be moved with the roots intact in the soil.

Seedage Methods for Fruits. Since most fruit trees are propagated by budding and graftage, seedage is of interest largely from the standpoint of rootstock production, and for the development of new varieties.

Seedlings of some fruits are grown with the intention that

they shall be budded or grafted in place before they are transplanted. Others are dug and sold as seedling rootstocks for replanting, either after they have been grafted or for budding and grafting at a future date.

The former practice is followed largely in the production of peach, plum, cherry, apricot, tung, pecan, and walnut rootstocks; the latter practice for apple, grape, pear, and, to some extent, citrus. In the North, fall planting is successful because of sufficient cold weather to break the rest period, if it is required by seed. In regions of mild winters, the seeds are subjected to necessary chilling in cold storage, and spring planting is more common.

Depth of planting and spacing vary with different kinds of fruits. Spacing is determined partly by the length of time the rootstocks are to remain in the nursery row before transplanting. Peach and other stone fruits remain in the nursery only one or two seasons. The seed can be planted from 4 to 6 inches apart in nursery rows and the resulting trees will have ample room for normal development until they are moved. Apple and pear, grown for transplanting after one year in the nursery row, can be given a much closer spacing by planting 15 to 18 seeds per foot. Pecan and walnut remain in the nursery 3 years and oftentimes longer. They require greater space for this growth and are customarily planted from 8 to 12 inches apart.

QUESTIONS

1. What are reasons for starting certain vegetable crops in special seedbeds while others are regularly field-planted?

2. What are the uses of horticultural plants that are grown from seed?

3. What factors influence the depth of planting and the spacing for seeds of fruits planted to produce rootstocks?

SUGGESTED REFERENCES

Stoutemyer, V. T., Albert W. Close and Claude Hope: Sphagnum Moss for Seed Germination, *U.S. Dept. Agr. Leaflet* 243, 1944.

Townsend, G. R.: Controlling Damping-off and Other Losses in Celery Seedbeds, *Florida Agr. Expt. Sta. Bull.* 397, 1944.

CHAPTER 8

Layerage

Stems that form roots while still attached to the parent plant are called *layers*, and the practice based on this phenomenon is known as *layerage*. In some plants artificial methods must be employed, while in others root formation occurs naturally. The rooting medium is usually soil, although other materials are used.

Uses. Layerage is a rather *certain* method of inducing rooting. Some plants that cannot be started satisfactorily from cuttings can be grown with relative ease from layers. A cutting, having been severed from the plant on which it grew, often does not remain alive until roots are formed. A layer, on the contrary, is supported by the parent plant indefinitely and, in the meantime, it is likely to develop roots.

Many plants produce *natural* layers freely and thus provide a ready source of new plants. This is true of the raspberry and strawberry and of certain forms of the blackberry and dewberry. In these plants the layers are produced by either runners or upright canes that, by arching, come in contact with the ground and develop roots. Other plants produce natural layers from the crown of the plant. The quince and chrysanthemum illustrate this behavior.

On a small scale, layerage may be used to good advantage, for the reason that the layers do not require the close attention as to watering, humidity, and temperature that cuttings require. Roses are sometimes grown from layers for this reason.

Objections to layerage are that it is a slow and cumbersome method of propagation; that it may interfere with cultivation; and that parent plants produce a limited number of new plants, so that a great number of stock plants must be provided. Despite

100

these disadvantages layerage is used quite commonly in the propagation of some plants, and certainly has a wide range of adaptation for the amateur gardener.

Simple Layers. Branches that have formed roots in one area only are called *simple* layers. Such layers are made by bending the branches to the ground and covering the portion just below the tip with 3 to 6 inches of soil. This practice is usually carried on in early spring, before growth has started. The tip of the shoot is left exposed, to form leaves and carry on the normal processes of the plant.

It is a common practice to injure the portion to be covered, by notching, cutting, girdling, or twisting. This practice destroys the phloem tissue, partially or completely, and retards the downward movement of food materials manufactured by the leaves of the exposed terminal portion. The result is an accumulation of plant food above the injured area, and such plant food is favorable to the development of roots by the layer. It is also considered that the injury checks the downward movement of hormones, and the concentration of these in the injured area stimulates root formation.

The season of the year for making layers varies with the species. With some the best results are obtained if layers are made in late winter or early spring; with others, late summer and fall seem to be the best seasons. The length of time during which layers are allowed to develop before they are severed from the parent plant likewise varies with different species. Many will make sufficient root and top growth during one season to permit them to be transplanted successfully to a new location; others require two seasons to develop a strong root system.

Many different kinds of plants can be grown from simple layers. In actual use, however, the method is restricted largely to very difficult species, and to plants grown for home use.

Tip Layers. A *tip* layer differs from a simple layer in that the tip is completely covered. Tip layers are used extensively in the propagation of some varieties of blackberries, dewberries, and raspberries. In starting new plants by this method the tips of branches are placed in the soil, pointing downward, to a depth of 2 to 3 inches, and covered. The soil is packed lightly to hold the branch securely in place. For the production of a large number of plants a shallow furrow may be plowed along the row a short distance from the plants, and all the available lateral tips laid in the furrow and covered. Tip layers of berries are best made in late summer. The covered portion will shortly become etiolated and fleshy. Adventive roots will develop in from 2 to 3 weeks, and the layer can then be dug, severed from the parent plant, and replanted in a permanent location. This can be done shortly after rooting occurs, but best results are obtained by allowing them to remain in place until the following spring, and replanting at that time. The rooted layer should be

Fig. 44. Natural tip layer of blackberry, showing growth of shoots at three nodes at *b*; new plants arising as suckers at *a*.

replanted with tip pointing upward since the stem will develop from the terminal bud.

Compound Layers. Long shoots that are alternately covered and exposed over their entire length are known as *compound* layers. They normally form roots at each node where they are covered and develop new shoots from buds at nodes that are not covered. When they have grown one season or more, the several layers are severed so as to provide a root system on the proximal portion of each layer and a top on the distal portion. The time of the year for making and for replanting compound layers is influenced by several factors. Normally they are made in late winter and early spring. The rooted layers may occasionally be replanted later in the same season; but more commonly they are allowed to grow one or two full seasons in order to develop a strong root system. Compound layerage is adapted to the propagation of the Muscadine grape. The natural production of "rosettes" and roots by the strawberry plant at each second node of the runners is similar to compound layerage.

Trench or Continuous Layers. This type of layer differs from the compound layer in that the branch is covered for its entire length instead of alternately. This method is adapted to the

FIG. 45. Blackberry shoots which have developed from a continuous layer.

propagation of own-rooted apple, pear, plum, cherry, and other plants needed for research investigations or other uses. It can also be used on Muscadine and other kinds of grapes that do not root well from cuttings.

Essentially, trench layerage consists of placing the main stem of a plant in a trench in a way that will permit young stems to develop from lateral buds and to form roots on the lower portions of these new stems. Plants that produce long vines can easily be bent to the ground. Others, like apple and pear, must be planted in horizontal position with the roots in proper contact with the soil and the main stem in the trench. Obviously, this would be practicable only with small whiplike plants.

In practice three methods are used in covering continuous layers. By one method, the layer is placed in an open trench. New shoots develop from lateral buds, and when they are about 6 inches high, soil is added to a depth of about 5 inches. Roots develop on the bases of the shoots that are covered with soil.

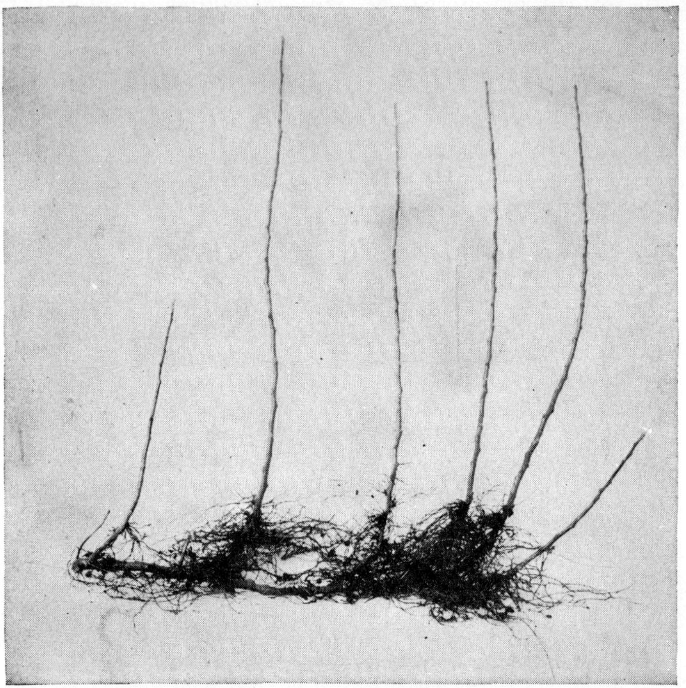

Fig. 46. Six rooted apple shoots removed from a layered plant. (*Courtesy of H. B. Tukey, N.Y. (Geneva) Agr. Expt. Sta.*)

By another practice, about 1 inch of fine soil is added when the layer is first placed in the trench. The new shoots push upward through this layer. As the shoots elongate, more soil is added around them until they are covered to a depth of 5 to 6 inches. The bases of shoots that develop when treated in this manner are etiolated, a condition favorable to ready root formation. By still a third practice, the layer is covered to a depth of about 3 inches with loose soil when it is made. The shoots push upward through this layer and develop roots from the etiolated portion of the stem below ground.

In every case, the roots arise adventively from the cambium layer of the new stems. The best season for making continuous layers is in late winter or early spring. The rooted plants are allowed to develop one full growing season before they are removed from the parent layer and replanted.

Mound or Stool Layers. This method is especially satisfactory for the rooting of apple and quince rootstocks and is used in preference to trench layerage when possible, as it involves less trouble and expense. A stock bed is established by setting young plants 2 feet apart in rows 3½ feet apart. The plants are headed back before growth starts and are allowed to grow for one season. The following winter the plants are cut back within 2 inches of the ground level, with the result that many new shoots arise from the base during the following season.

In the case of apples, which root freely from these new shoots, the stools are allowed to remain uncovered during the early part of the growing season. The greatest number of shoots are produced in this way; after they are formed and have reached the height of 8 inches they are mounded with 5 to 6 inches of soil. Mounding should be done with moist soil, which should be placed from the center outward, in order to bend the shoots out and give them better spacing. This spacing seems to give a better rooting, especially with vigorous shoots.

When plums are being grown, the procedure is modified and the plants mounded before the new shoots appear. This practice results in the formation of fewer new shoots than the other method, but the shoots that are produced are etiolated and form roots better than those that are produced before mounding. This applies not only to shoots from stools and layers but also to stems used for cuttings, from which better rooting is obtained when their bases have been etiolated during growth.

After the plants have been mounded by either method in early spring, they are allowed to grow during the rest of the season, and roots will form on the new shoots along the covered portions of the stems. In early winter the rooted shoots are removed and planted in the nursery row. These plants are set at a depth of about 6 inches. They will be ready to bud during the summer of the following year, or they may be grafted at the end of one season in the nursery.

The chrysanthemum forms natural mound layers from the overwintering crown at the beginning of each new growing season. These develop into new plants when they are detached and planted out separately. Quince and Japanese flowering quince have habits of growth that permit them to be propagated from natural layers from the crown of the plant. Varieties of

FIG. 47. Rooted shoots of layered apple, separated and ready to be used as lining-out stock. (*Courtesy of H. B. Tukey, N.Y. (Geneva) Agr. Expt. Sta.*

currants and gooseberries that do not grow readily from cuttings are frequently grown from mound layers.

Air Layers. A method used to root branches of upright growing plants that do not sprout or sucker readily is known as *air layerage*. Chinese layerage and pot layerage are other names for the same method.

The stem is first injured by slicing, notching, ringing, or binding. Care must be exercised not to injure it sufficiently as

Fig. 48. Chrysanthemum plant, showing formation of new plants as natural layers from the crown of an old plant.

Fig. 49. Air layer of *Ficus pandurata* showing plastic film enclosing moss at X and the same layer after it had formed roots in the moss, shown at Z, and had been detached for planting.

to cause breakage or death of the layer. This can be effected easily by binding with copper wire wrapped tightly about the stem, and it has the same effect on rooting as the other treatments. It is common practice to apply a coating of one of the concentrated hormone dusts to the area where roots are to form.

This is particularly helpful in the rooting of difficult species. The injured area on the stem is then covered with a handful of moist sphagnum, which is tied in place and kept continuously moist by sprinkling. By another method the moss is covered with plastic strips or sheets to retain a constant level of moisture in the area where new roots are to be formed. The film to be used must retain moisture and at the same time be permeable to both oxygen and carbon dioxide. A plastic known as polyethylene is a material that meets these requirements. It is available in sheets of various sizes, and is used commonly with air layers. This material is wrapped closely about the moist moss, drawn in tightly at the top and bottom, and tied securely with rubber bands, tape, or string. The film will keep the moss moist for a prolonged period, and the layer does not require further attention until it is rooted.

The air layer may be made on stems of one-year wood or older; but the older branches are often slower to root and they become reestablished less readily when moved.

The time required for air layers to develop roots varies with the species of plant being propagated, from a few weeks to a year or longer. When sufficient roots have developed, the layer is severed from the parent plant and replanted in a permanent location or container. When these new plants are detached and transplanted, best results will be obtained if they are kept in a cool, shaded location until they have become established and have renewed their growth.

Air layerage is used exclusively in the propagation of named varieties of lychee trees. It is also used successfully in the propagation of bougainvillaeas, hybrid crotons, hybrid hibiscus, dracaenas, panduratas, and many other kinds of ornamental plants. It is frequently used as a novelty method of propagation on plants that can readily be propagated by other methods.

Layerage as a Preliminary Treatment. Layerage is sometimes used as a preliminary treatment for the rooting of cuttings. It frequently happens that the layered parts do not form satisfactory root systems in the first season, but the stem pieces with small roots may be separated from the plant and treated as cuttings. Another use of preliminary layerage is for the purpose of etiolating the stem so as to induce rooting by cuttings. Work done by several investigators in England shows that the etiolated portion of the young stem forms a superficial starch sheath and that roots form more readily when this condition occurs.

QUESTIONS

1. What is layerage? List the different types, and prepare sketches illustrating each.

2. Under what condition is it used in preference to some other method?

3. What influence does notching or girdling have on root formation?

4. What is etiolation? List the different kinds of layers in which the shoots are etiolated as they develop into layers, and the kinds in which rooting occurs from a nonetiolated shoot.

5. What is the tissue from which roots arise in layers?

6. What are the advantages of using plastic covers for air layers? What is the name of one that is commonly used? What are its particular merits?

7. Outline appropriate seasons for making and for replanting a layer made by each one of the different methods.

SUGGESTED REFERENCES

Baker, R. E., and H. M. Butterfield: Commercial Bushberry Growing in California, *Calif. Agr. Ext. Serv. Circ.* 169, 1951.

Creech, John L.: Layering, *National Horticultural Magazine* (Washington, D.C.), **33**:37–43, 1954.

Darrow, George M.: Growing Erect and Trailing Blackberries, *U.S. Dept. Agr. Farmers' Bull.* 1955, 1948.

Flint, W. P.: Bramble Fruits, *Ill. Agr. Ext. Serv. Circ.* 427, 1935.

Gardner, F. E.: The Vegetative Propagation of Plants, *Maryland Agr. Expt. Sta. Bull.* 335, 1932.

Hansen, C. J., and E. R. Eggers: Propagation of Fruit Plants, *Calif. Agr. Ext. Serv. Circ.* 96, 1936.

Knight, R. C., J. Amos, R. G. Hatton, and A. W. Witt: The Vegetative Propagation of Fruit-tree Root Stocks, *East Malling Research Sta. (Kent, Eng.) Ann. Rept.* (14th and 15th years), II, Supplement, pp. 18–19, 1928.

Stahl, J. L.: Propagation of Deciduous Fruits, *Calif. Agr. Expt. Sta. Circ.* 294, 1925.

Talbert, J. T.: Plant Propagation by Seedage, Cuttage, Layerage, and Separation, *Missouri Agr. Expt. Sta. Circ.* 191, 1936.

Watkins, J. V.: Propagation of Ornamental Plants, *Florida Agr. Ext. Serv. Bull.* 150, 1952.

CHAPTER 9

Cuttage

Cuttage is the process of propagating plants by the use of vegetative parts that, when placed under suitable conditions, will develop into complete plants. It differs from layerage in that the parts used are detached from the parent plant before they have an opportunity to develop roots. With species of plants that strike roots readily, cuttage is a cheap and convenient mode of propagation. It is used extensively in the propagation of ornamental plants, including deciduous types, broad-leaved evergreens, and coniferous forms. Some fruits, such as grapes and figs, have been propagated in this manner since ancient time, and more recently there has been considerable progress in the rooting of other fruit plants, such as the Bruce plum. In the majority of cases, however, the rooting of fruit-tree species is of more importance in the production of uniform stocks for budding or grafting.

Classes of Cuttings. Plant parts used in making cuttings fall into four groups: roots, leaves, stems, and modified stems (tubers, rhizomes, and similar structures). Theoretically, all plants that have primary meristems are capable of being propagated by cuttings. All plants cannot profitably be increased by this means, however, and only practical experience has made it possible to distinguish between species that can be propagated from cuttings and those that cannot.

Root Cuttings. As a rule, plants that naturally produce suckers freely can be propagated easily by root cuttings. Some species of plants that root rarely or not at all from stem cuttings can be reproduced by this means. Persimmon, pear, pecan, apple, and plum are of this class. They may be started by root cuttings, but other methods are considered more economical and are in general use. Sweetpotato and horseradish are propagated commercially by root cuttings, and blackberries and raspberries may be propagated successfully by this method. It should be borne in mind, however, that a root cutting will perpetuate the part of the plant from which it was secured. A root taken from below the union of a budded or grafted tree reproduces the

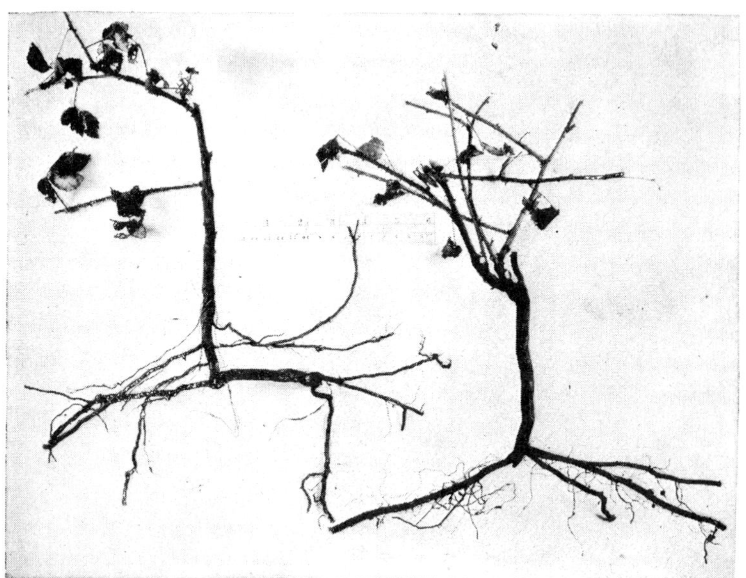

Fig. 50. Blackberry plants that have grown one season from root cuttings; one on left is from a root cutting planted horizontally and one on right is from a root cutting planted vertically.

seedling stock of unknown bearing quality rather than the standard top.

The technique of making root cuttings varies widely with different species. They are customarily made from roots that are not smaller than ¼ inch in diameter, which are cut in lengths of 2 to 6 inches. They may be made early in winter, stored in sand, and allowed to callus. They are then planted out in the open the following spring. By another practice the cuttings are started in early winter in greenhouses or hotbeds and transplanted to the open after they have made top growth and formed new roots; such plants are usually large enough to be transplanted by spring. Root cuttings are also planted directly in the field in the spring, without preliminary treatment. They may be planted in either a horizontal or vertical position; if planted vertically, the end that was nearest the crown of the parent plant should be uppermost. New shoots develop from root cuttings

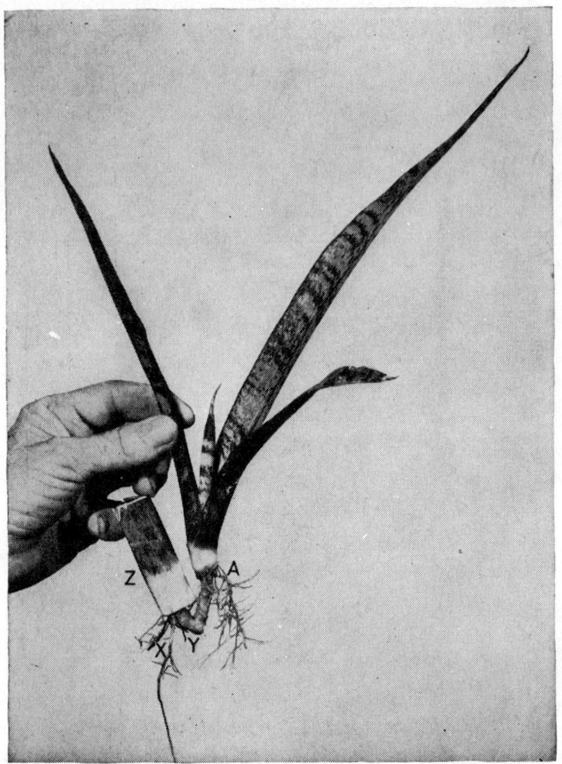

FIG. 51. Snake plant (*Sansevieria* sp.) leaf cutting shown at Z. It produced roots at X, and also the rhizomelike stem at Y, which developed into the new plant with roots shown at A.

from adventitious buds, and new branch roots form adventively in the cambium, either from the old root part used as a cutting or from the base of new shoots that develop from below ground.

Leaf Cuttings. Many plants with thick or fleshy leaves can be propagated by leaf cuttings. Thin-textured leaves usually dry up before rooting can take place. Practices vary in the actual preparation and planting of leaf cuttings. In some cases the leaf is detached from the parent plant and planted vertically in a suitable medium with the petiole and about one-half of the leaf covered. Adventive roots and shoots both develop at the base, usually from the petiole. These arise normally from parenchymatous tissue closely associated with the vascular cambium, and also in the primary rays. The lemon is an example

FIG. 52. Peperomia leaf cutting Z, produced roots from petiole at *X*, and new shoot.

of a plant that can be grown from leaf cuttings planted in this manner.

By another practice, the leaf is placed flat on sand in a propagating bed, cut transversely across the center vein, and then covered lightly with sand. Adventive shoots will develop where veins were severed, and adventive roots will develop from the bases of the new shoots. Species of *Bryophyllum* can be grown from leaf cuttings made in this manner.

Leaves of snake plant (*Sansevieria*) when cut into several segments and planted separately, with the basal portion of each inserted into the rooting medium, will develop into new plants. Adventive roots arise near the lower cut surface from cells adjacent to the suberized layer. New stems arise adventively somewhat later from callus tissue that forms at the base.

Fig. 53. Cuttings of French crab seedlings. 1 and 2, in sand, unrooted; 3 and 4, in sand and peat, early in season—basal roots; 5 and 6, in sand and peat, later in season—basal and nodal roots.

In general, roots will develop more readily than shoots from leaf cuttings. Many plants, such as the rubber plant, will form roots, but rarely tops, from leaf cuttings. This difficulty is avoided by using leaf-heel cuttings, consisting of a leaf with a small sliver of the stem and the axillary bud; and also by using leaf-mallet cuttings, consisting of a segment of the stem and the axillary bud. These are in reality miniature stem cuttings. They are planted with the heel or mallet and the base of the leaf covered lightly. The axillary bud develops into a new stem, and new roots form adventively from the heel or mallet and also from the base of the new shoot. The leaf supports the cutting and young plant and remains attached for a prolonged period, but it does not become a part of the new plant.

Stem Cuttings. These are made from herbaceous plants, such as those frequently grown in greenhouses, and from woody plants, which are usually grown in the open. Cuttings of woody plants may be classed as semihardwood, or softwood, and hardwood, depending upon the stage of growth of the wood used.

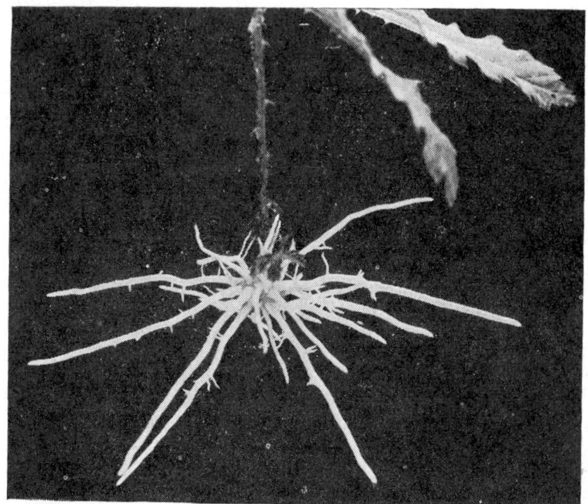

Fig. 54. Hardwood stem cutting of blackberry, with good root development.

Herbaceous Cuttings. These are made mostly of greenhouse plants that are herbaceous in type. Cuttings of such material are usually soft, tender, and succulent; they require special attention with regard to temperature and moisture to prevent wilting. Under favorable conditions they root satisfactorily in a relatively short time. Examples of plants that may be propagated by herbaceous cuttings are geranium, coleus, petunia, alternanthera, chrysanthemum, tomato, and sweetpotato.

Semihardwood Cuttings. Stem cuttings of trees and shrubs that are made from current-season shoots are known as *semihardwood,* or *softwood,* cuttings. In practice they are made 3 to 6 inches long. Cuttings that are made so as to include terminals of growing shoots are generally preferred, though those made from parts below the terminal are satisfactory. Shoots that snap clean when broken are considered to be in ideal condition for use as semihardwood cuttings. The leaves are removed from the basal portion, but those near the terminal are left.

Semihardwood cuttings are succulent and tender; for this reason it is important that they be handled so as to prevent wilting after they are cut and before they are planted. The presence of leaves causes a high rate of transpiration, which makes this difficult. Best results may be secured by cutting them during a cool part of the day, preferably in the early morning, while the material is turgid. They should then be wrapped in moist cloth or moss until planted. Such cuttings are usually started in specially prepared beds in a greenhouse, hotbed, or cold frame; some, such as blueberry, are sometimes started outdoors in special beds.

In addition to cool temperature, shade, and a high humidity, which are essential factors for good results with semihardwood cuttings, bottom heat may also be supplied in order to provide more desirable conditions for rooting. Manure is frequently used for this purpose, or the beds may be heated with flues, hot water, or electric heating elements. Shade may be provided by stretching domestic cloth at a height of 3 to 4 feet above the bed, or the glass of the greenhouse may be sprayed with lime whitewash to provide the same effect. On a small scale, cuttings may be planted in shallow boxes or flats placed in a shaded location. The cuttings and adjacent areas are sprayed with water several times a day to keep the cuttings from wilting.

Hardwood Cuttings. These are made from a wide variety of plants, including deciduous types, conifers, and broad-leaved evergreens.

Cuttings of *deciduous* plants are taken during the dormant season. Those of some plants are taken in the fall, packed in moist insulating material, and stored at a temperature of 40°F. or less. These cuttings are usually placed in the bed about midwinter. While in storage they may have formed callus at each end; this, however, is not essential to rooting. Instead of the procedure just outlined, cuttings of some deciduous plants are taken and planted in late winter, shortly before they would normally resume growth.

Deciduous cuttings may be made from 4 to 12 inches long, depending on the kind of plant. Usually one-year-old wood is used, but the older wood also may be rooted. It is a customary practice to make the top cut slightly above a node and the lower cut slightly below a node. Various kinds of cuttings show different responses with regard to the point of origin of roots; but the denser tissue in the vicinity of the node is thought to be of value in preventing drying out or decay of the wood. Deciduous hardwood cuttings are not highly perishable, but they should be protected at all times to prevent them from becoming dry.

Many species of plants may be propagated by hardwood cuttings set directly in the nursery row. Grape, fig, and rose are commonly propagated in this manner. Rooting is determined partly by the type of soil in which they are planted; sandy loam soil that is well drained is preferred. In order to ensure good aeration, cuttings are frequently planted on high beds. In a heavy clay soil in Oregon, grape cuttings rooted well when set in holes made with an iron bar and filled with sand.

Hardwood cuttings include also those made from mature wood of *conifers*. Cuttings of such plants are made 4 to 6 inches long with foliage removed from the lower portion of the stem. As the cuttings form roots, new shoots also form, and this top growth is an indication that the cutting is ready to be moved. The customary procedure is to pot the rooted plants and grow them in the pots for one season before moving them to the field. Some of the arborvitaes root within 2 or 3 months; junipers frequently require 6 months or even longer.

Several *broad-leaved evergreen* plants are grown from hardwood cuttings. The cuttings of certain citrus species, for example, are made 4 to 7 inches long with five or six nodes, from mature terminal growth. Leaves are removed from the lower part of the stem, but two or more are left at the top. As with other types of cuttings, it is important that cutting material be obtained from healthy, vigorous-growing trees. Orange, grapefruit, lemon, American holly, yaupon, and several species of Ligustrum are examples of broad-leaved evergreens that may be propagated by hardwood cuttings.

Origin of Roots in Hardwood Cuttings. Roots that develop in hardwood cuttings arise largely in the cambium layer. The conditions that are favorable to root formation are also favorable to callusing. The two processes frequently develop simultaneously, but aside from the occasional roots that actually arise from the callus tissue, there is no direct relationship between callusing and root development. Some plants tend to form roots only at certain locations on the cutting, some at the base only, others at nodes along the stem, and still others at nodes and internodes. Cuttings made at one season may form roots in one

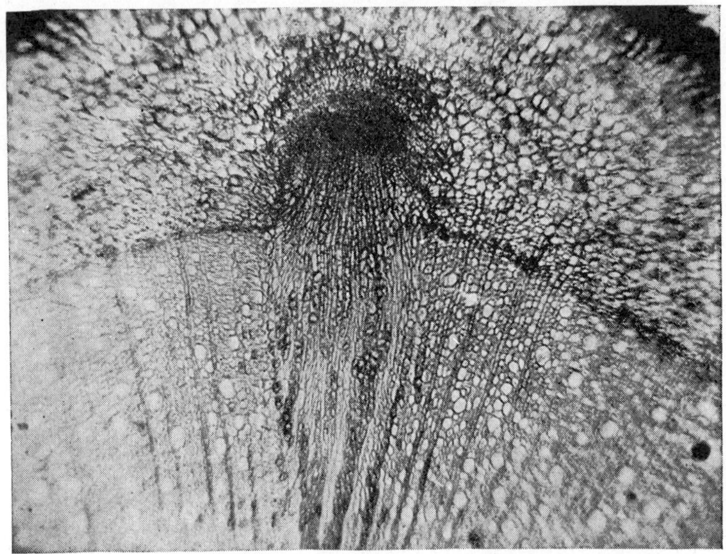

FIG. 55. Root primordia in quince stem.

manner from the standpoint of distribution, while the same kinds made at another season will form them according to an entirely different pattern.

The adventive roots of hardwood cuttings may arise in the cambium, usually in the rays, just prior to their emergence; or they may arise and exist in the cambium layer as preformed roots, known as *root primordia,* for an indefinite period prior to their emergence. These root primordia occur in the willow, certain cotoneasters, some species of gooseberries, and at least one variety of apple, the Springdale. Plants that have these preformed root initials usually grow readily from hardwood cuttings. Most woody plants that have been studied do not possess them, however, and of these some develop roots in normal fashion and grow readily from hardwood cuttings, while others do not.

Cuttings of Modified Stems. Material of this type is usually handled as bulbs and bulblike structures, and it will accordingly be discussed in connection with bulbs.

External Factors Influencing Root Formation. The rooting and growth of cuttings depends upon certain external or environ-

Fig. 56. Root primordia at two points in willow stem.

mental factors which represent treatments that are applied just before the cuttings are set in the bed, or the conditions to which the cuttings are subjected in the bed.

Media. Several different media are used in the propagating bed into which cuttings are to be planted. Any medium used should be loose and easily worked to facilitate planting of the cuttings, and particularly the removal of the cuttings with little damage to roots; it should be fairly retentive of moisture, and yet well drained; it should be free from fungi and bacteria which will attack cuttings; and it should be freely available at reasonable cost.

Clean sharp sand is used more commonly than any other

material. Ordinary building sand is nearly always satisfactory for this type of work. Best results will be obtained if it is screened properly to remove foreign material and is washed. Decaying organic matter in the sand is very objectionable, since it promotes the development of fungi and bacteria, which in turn may cause the cutting to die before root formation can take place.

Sand used in the cutting bed should be changed regularly after the plants are removed or should be sterilized effectively before being used again. This is especially important because in the transplanting of rooted cuttings a considerable portion of the root system may be broken and left in the soil. Dead cuttings are also frequently left in the bed too long, so that ready sources of infection are provided for the next lot of cuttings.

Loose sandy loam soil with good drainage is most suitable for cuttings planted directly into the nursery, such as fig, grape, and rose.

Acid peat has been used successfully as a rooting medium for cuttings in recent years. Peat is composed largely of partly decomposed organic material. It is normally brown in color, light and granular in texture, and acid in its reaction. It has a high water-holding capacity; saturated peat contains over three times its weight of water but is well aerated even when saturated. Aeration and retention of moisture are two important requirements of a rooting medium. The acid reaction of the peat is considered to be beneficial, or even necessary, for some cuttings that root more satisfactorily in it than in sand; it serves also to prevent bacterial decomposition. Cuttings of plants that root poorly when placed in sand often root satisfactorily in a mixture of equal volumes of sand and peat. The superiority of this medium over sand for certain species is probably due to improved aeration and increased water-holding capacity.

Vermiculite is a comparatively new material that is a satisfactory medium for the rooting of some cuttings. It is made up of very thin, flat particles of mica, which has been subjected to very high temperature. It is very retentive of moisture, owing to the great surface area of the thin particles. Watering the cutting bed of vermiculite must be done with care, and mist humidification is usually avoided to prevent overwatering. Perlite is another granular mineral compound that is used for the cutting bed. It is sterile and similar to sand in texture. Various other materials are used, such as coconut fiber, sawdust, and sphagnum moss.

Temperature. Control of temperature is a very important factor in the rooting of cuttings. Though high temperature is favorable for the rooting of some species, it stimulates a high rate of transpiration, particularly for herbaceous and semihardwood cuttings, which may result in wilting and death unless a high humidity is maintained.

In the case of hardwood cuttings, planted in the bed in winter or early spring, the primary consideration is to induce root activity before shoot growth occurs. For this reason it is desirable to provide bottom heat, so that the bed itself is 5 to 10°F. warmer than the surrounding air. This is accomplished by placing heating pipes or electric heating elements below the surface of the rooting medium in the bed. Heating pipes placed underneath a raised bed provide the same effect. Root formation may occur over a wide range of temperature, but a soil temperature of 65 to 70°F. gives satisfactory results with many plants.

Humidity. A high degree of humidity should be maintained in the cutting bed in order to prevent drying and death of the cuttings before they have opportunity to root. This is especially important for herbaceous, softwood, and evergreen cuttings. Frequent sprinkling of walls, walks, and beds in the greenhouse is necessary in order to keep the cuttings from drying and wilting, particularly under arid conditions.

Equipment is available for automatically maintaining constant humidity for propagation beds. So-called mist humidification is a means of maintaining high humidity and also for sprinkling the cuttings, both of which are helpful in preserving the turgor of the cuttings.

In hotbeds, cold frames, or other propagating structures, glass sash or other types of covers are used to prevent water loss. Under such conditions of high humidity, diseases of all kinds find a favorable condition for rapid spread. Careful sanitation of the cutting bed is one of the essential considerations to keep down such diseases.

In sprinkling to maintain a high humidity caution must be exercised to prevent overwatering of the cutting bed. A well-drained bed and well-aerated rooting medium will help preclude this.

Chemical Treatments. Many kinds of chemicals have been used in efforts to induce root formation in species difficult to propagate or to increase the number and extent of roots in others that develop slowly. Early work with dilute solutions of potassium permanganate on privet cuttings attracted attention to this method as a possible aid in stimulating root formation. Dilute solutions of vinegar and of cane sugar were also used successfully with some types of plants. Many other materials, appar-

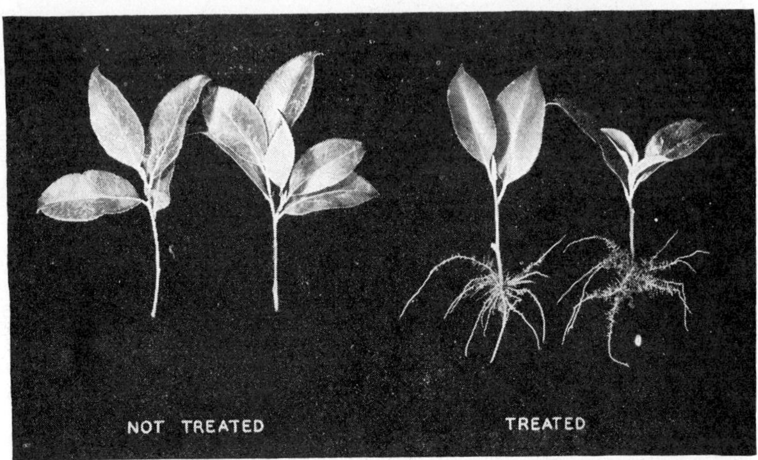

Fig. 57. Camellia cuttings. Photograph shows characteristic response of these which are not treated in comparison with those which are treated with Hormodin, a preparation of indolebutyric acid. Picture shows the cuttings as they appeared 60 days after they were placed in the rooting medium. (*Courtesy of Hitchcock and Zimmerman, Boyce Thompson Institute for Plant Research, Inc.*)

ently chosen at random, were used in the hope of inducing the desired stimulation for root development. Many of these treatments gave favorable results with some plants, but in other cases the percentage of rooting was not increased above that of the checks, and in still others rooting was definitely retarded or entirely prevented.

More recently certain growth regulators or hormones, derived from plant tissue or produced synthetically, have been used to stimulate plant growth and especially root formation in cuttings. The most widely used of these are indole-3-acetic (IA), indole-3-butyric (IB), and napthalene-1-acetic (NA) acids. In addition, ethylene, acetylene, propylene, and carbon monoxide gases stimulate development of roots of cuttings of certain species.

Several methods of introducing the auxins in the cutting have been used. One method is to soak the bases of the cuttings for a period up to 24 hours in a solution that contains from 5 to 50 and sometimes more, milligrams per 1,000 cubic centimeters of water. A variation is to soak the entire cutting instead of the base only. Treatment under vacuum results in better penetration

FIG. 58. Carnation cuttings, photographed three weeks after planting in the cutting bed. Left, planted without treatment; right, dipped in Hormodin powder before being planted. (*Courtesy of Hitchcock and Zimmerman, Boyce Thompson Institute for Plant Research, Inc.*)

of the auxin. Another method is to dip the bases of the cuttings in a concentrated solution of the chemical, prepared by adding from 5 to 20 milligrams of the growth regulator to 1 cubic centimeter of 50 per cent methyl or ethyl alcohol. This method is effective and is preferred by some propagators because of its adaptation to practical use. Growth regulators are also applied to cuttings in powder form. The bases of the cuttings are moistened, and then dipped in a mixture of from 5 to 15 parts of the growth regulator and 1,000 parts of talc.

The best stimulation of root growth is usually obtained from concentrations slightly below the toxic level. Within the range of concentrations suggested, the strongest would be applicable to hardwood cuttings, the intermediate range to semihardwood, or certain types of evergreen plants, and the lowest range to herbaceous plants.

Briefly, growth regulators promote rapid and heavier rooting of cuttings which ordinarily root well. They are less helpful with difficult species and in the rooting of cuttings of deciduous fruit and nut trees.

Mechanical Treatments. Mechanical treatments of various kinds have been used to stimulate root formation. Some of these treatments are used on the plant before the cuttings are made, with the result that internal or structural changes are induced. For this reason they are included under the internal factors.

The presence of leaves on cuttings provides a favorable influence on the rooting of herbaceous, semihardwood, and evergreen hardwood cuttings. Since they are the primary photosynthetic part of the plant and since rooting is enhanced by a high level of carbohydrates, it would be expected that, within limits, the rooting response is proportionate to the leaf area. Leaves also have additional influences on the rooting of cuttings. They provide auxins and other organic materials essential for growth. Evidence that leaves do have a direct influence on rooting is found in these observations: In the case of citrus cuttings, the removal of the terminal half of each leaf retards root formation and reduces the total amount of roots produced. If leaves are removed from all but one side of a cutting, roots form primarily on the side with leaves. Cuttings with entire leaves root better than those with the same total leaf area of half-leaves. The main reasons for reducing the leaf area of cuttings is to reduce the loss of moisture and to facilitate planting in the propagating bench. Leafy cuttings planted in complete darkness do not root properly because the stored food supply becomes depleted in respiration and growth and is not replaced.

The rooting response of some plants is determined partly by the position of the basal cut. Some plants root best if the basal cut is made so that a node is left at the base of the cutting; a few root best if an internode occurs at the base; and still other plants root best if the basal cut is made slightly below a node.

A wound made at the base of the cutting oftentimes causes better root formation. This is done by cutting a sliver of bark and wood, $\frac{3}{4}$ to $1\frac{1}{2}$ inches long, from the side of the cutting at the base, or by merely slitting the bark on one or two sides. These treatments increase the area from which roots may be expected to form, and roots often develop along the margins of the wound.

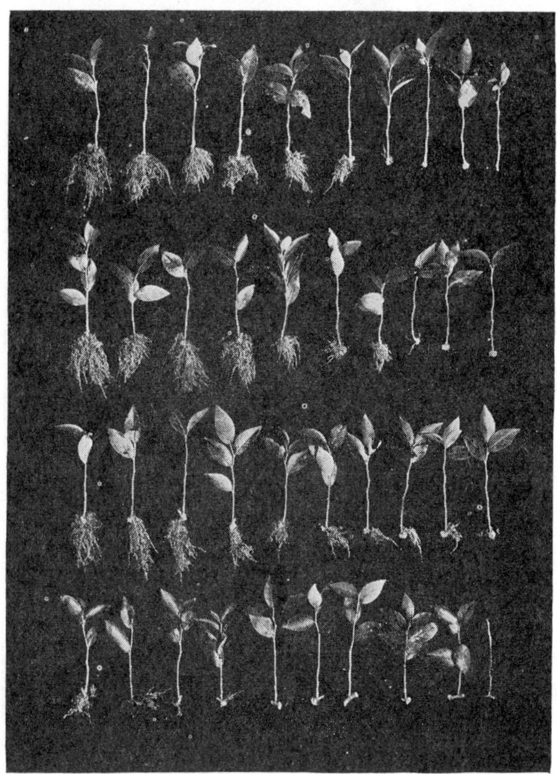

FIG. 59. The difference in rooting of highbush blueberry (*Vaccinium corym-bosum*) from four types of softwood cuttings. Top, cut made above base; second row, cut made at base; third row, heel cuttings; bottom row, mallet cuttings. (*Courtesy of Hitchcock and Zimmerman, Boyce Thompson Institute for Plant Research, Inc.*)

The type of wood at the base has an important influence on the rooting of cuttings of some plants. Cuttings may be made of current-season growth or green wood, with the basal cuts made in one of the following ways: (1) cuts made above the base of the current season's growth producing a *terminal* or *subterminal* cutting; (2) cuts made at the base of current season's growth producing a *basal cutting;* (3) cuts made to include a heel of one-year-old wood forming a *heel cutting;* and (4) cuts made to include a mallet of one-year wood forming a *mallet cutting.*

Fig. 60. The difference in rooting of a species of plum (*Prunus tomentosa*) from four types of softwood cuttings. Top row, cut made above base; second row, cut made at base; third row, heel cuttings; bottom row, mallet cuttings. (*Courtesy of Hitchcock and Zimmerman, Boyce Thompson Institute for Plant Research, Inc.*)

Internal or Structural Factors. Internal or structural factors represent conditions within the cutting, which may influence its ability to form roots and develop into a plant. Such conditions may be affected by treatments to which the cuttings are subjected some time before they are removed from the plant. These treatments differ from external treatments in that they are designed to induce some change in the chemical composition or structure of the material before it is planted in the cutting bed.

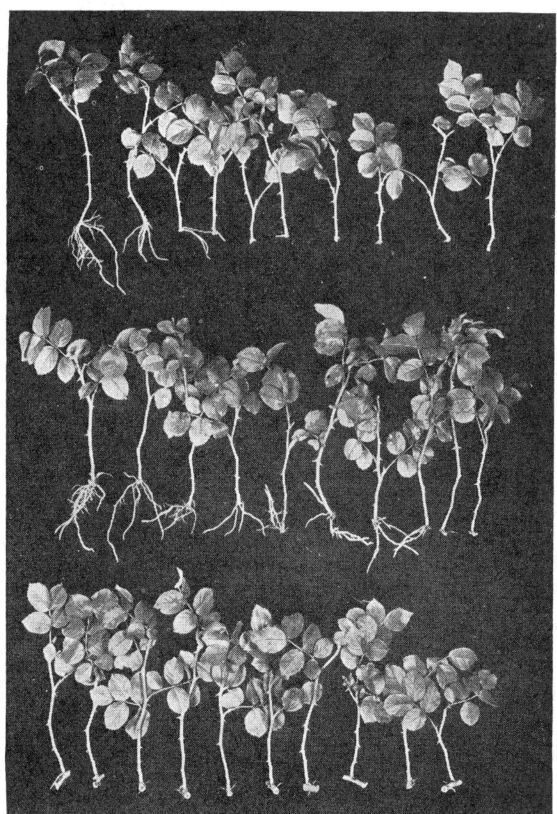

FIG. 61. The difference in rooting of American Pillar rose from three different types of cuttings. Top row, cut made above base of shoot; middle row, cut made so as to include a bud of last season's wood; bottom row, mallet cuttings. (*Courtesy of Hitchcock and Zimmerman, Boyce Thompson Institute for Plant Research, Inc.*)

Stored Food. Two general requirements are necessary in the formation and growth of roots on cuttings—the plant must have the capacity to develop root and top growth, and energy must be supplied for these processes. It has been shown repeatedly that the available carbohydrates and nitrogen markedly affect the rooting of cuttings.

In California, cuttings of Sultanina grape were sorted into three classes on the basis of their starch content. The freshly cut ends of the cuttings were dipped in a solution of iodine in potassium

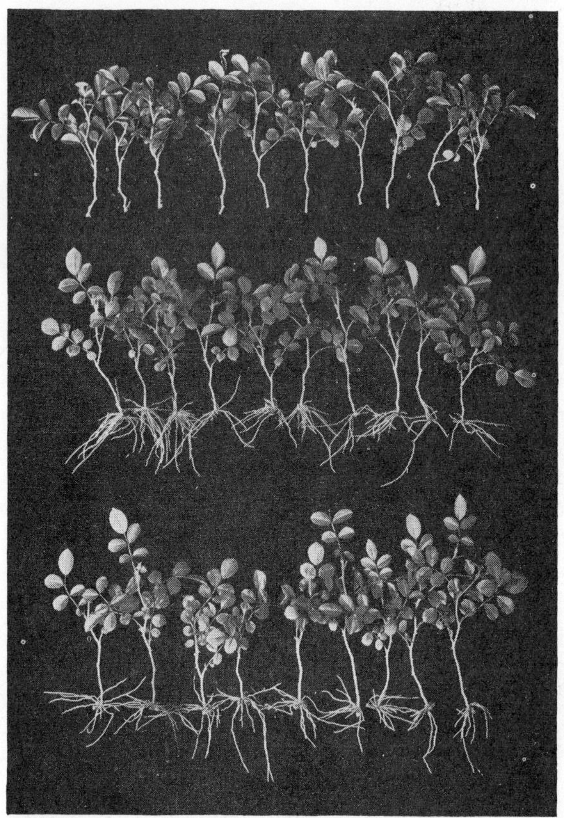

Fig. 62. The difference in rooting of Dorothy Perkins rose from three different types of cuttings. Top row, cut made above base of shoot; middle row, cut made so as to include a bud of last season's wood; bottom row, mallet cuttings. (*Courtesy of Hitchcock and Zimmerman, Boyce Thompson Institute for Plant Research, Inc.*)

iodide, and the intensity of the staining in wood outside the medullary rays was used as an indication of the amount of starch present. Cuttings that showed the deepest stain rooted 62 per cent and formed good roots; the intermediate group, 35 per cent with moderate roots, and the low-starch group, 17 per cent with very poor root systems.

Shoots of stock plants from which cuttings are to be taken are sometimes girdled in order to influence the amount of stored food that the cuttings will contain. The girdle is made at the point on the stem which will be the base of the cutting. The resulting swelling above the girdle is accompanied by an accumulation of stored food at this point and also naturally occurring auxins that move from the top of the plant toward the base. The girdling of the shoot is done during the growing season as soon as length growth ceases, and the material is then removed for cuttings during the following dormant period. The additional amount of reserve food accumulated at the base of the cutting is of value in promoting root formation; the method, however, would be practiced only for plants that are difficult to propagate.

If nitrogen is plentiful, and carbohydrates are low, growth of shoots is stimulated, but rooting is slight. Cuttings from plants that have made normally vigorous growth and hence have a carbohydrate accumulation in excess of inorganic nitrogen are more likely to root properly.

Age and Maturity of the Tissue. There is a definite relationship between the maturity of the tissues of a cutting and the readiness with which it forms roots. If the cutting is soft and immature, it becomes weakened more readily from transpiration and more susceptible to decay; and if the tissue is old and mature, a longer period of time is required for satisfactory rooting. In actual practice propagators learn that certain kinds of plants can be grown best from cuttings representing a certain stage of maturity, and that other kinds can be grown best from those representing an entirely different stage. Specifically, some plants grow best from semihardwood cuttings and show differences in the response of terminal and subterminal; others grow best from basal cuttings with tissues that are more mature; there are also plants which root more readily from heel and mallet cuttings in which second-year wood is included. Grape and certain plums grow readily from one-year-old-hardwood cuttings. The olive is propagated by means of "truncheon," which consist of wood that is several years old. It has been shown that hardwood grape cuttings taken from the middle and basal region of a stem normally root better and produce more vigorous plants than those from near the terminal; they possess more carbohydrates and less nitrogen than stem tips.

Callusing. Callus formation at the basal end of the cutting was at one time considered to be a vital factor in the rooting of hardwood cuttings. More recently it has been accepted that it does not play an important part in root formation. Some few cases have been observed where roots originated in the callus tissue, but that is uncommon. Callus formation may be of benefit in sealing the end of the cutting and preventing decay. Callused cuttings

FIG. 63. Callus and root formation on a cutting. (*Courtesy of Homer E. Rea, Tex. Agr. Expt. Sta.*)

also respond more readily to chemicals used to aid in root formation than those not callused.

Etiolation. Parts of shoots not containing chlorophyll are said to be *etiolated,* and this condition is regarded as being favorable to root formation. Some investigators have attributed better rooting to the formation of an endodermis, as in roots. Etiolation may be produced by wrapping stems with tape or by covering with soil. The exclusion of light causes chlorophyll to disappear. In some cases shoots are caused to develop in darkness, by mounding with soil: chlorophyll never develops. Stems arising from below the ground, as in mound or continuous layerage, are etiolated.

QUESTIONS

1. Define cuttage.
2. Give some advantages of propagating plants by cuttage.
3. Why are own-rooted plants sometimes not desirable?
4. Name the different plant parts from which cuttings are made.
5. What tissue gives rise to new roots when the following kinds of cuttings are made: Leaf? Stem? Root?
6. What are preformed roots?
7. What factors determine the number of leaves to be left on herbacious or semihardwood cuttings?
8. Outline various treatments that influence the rooting of cuttings.
9. Outline briefly the influence of these on rooting: Leaves? Stored food? Auxins? Callus? Etiolation?

SUGGESTED REFERENCES

Curtis, O. F.: Stimulation of Root Growth in Cuttings by Treatment with Chemical Compounds, *Cornell Univ. Agr. Expt. Sta. Mem.* 14, 1918.

Gardner, R. J.: Propagation by Cuttings and Layers: Recent Work and its Application, with Special Reference to Pome and Stone Fruits, *Imp. Bur. Hort. Plantation Crops,* East Malling, Kent, England, *Tech. Commun.* 14, 1942.

Halma, F. F.: The propagation of Citrus by Cuttings, *Hilgardia,* **6:** 131–157, 1931.

Pease, Roger W., Earl H. Tryon, and W. W. Steiner: Rooting American Holly from Cuttings—Cold Frame Method, *West Va. Agr. Expt. Sta. Circ.* 87, 1953.

Snyder, William E.: The Rooting of Leafy Stem Cuttings, *Natl. Hort. Mag.,* vol 33, 1953.

Tukey, H. B., and Karl D. Brase: The Propagation of Multiflora Rootstocks for Roses by Softwood Cuttings, *N.Y. (Geneva) Agr. Expt. Sta. Bull.* 598, 1931.

Winkler, A. J.: Some Factors Influencing the Rooting of Vine Cuttings, *Hilgardia,* **2:**330–349, 1927.

CHAPTER 10

Bulbs and Other Modified Structures

Most horticultural plants are characterized by normal roots, stems, leaves, flowers and fruits. There are, however, certain plants in which one or more of these parts have become highly modified. These plants constitute some of the most valuable and interesting flowering and vegetable plants.

Classification. The various types of bulbs and other modified structures may conveniently be included in the following classification:

1. Bulbs
   a. Layered or tunicate. Examples: onion, garlic, narcissus, hyacinth, and tulip.
   b. Scaly. Example: lily.
2. Corm, or solid bulb. Examples: gladiola and crocus.
3. Rootstocks
   a. Rhizome. Examples: canna, banana, bamboo, and asparagus.
   b. Pip. Examples: Lily of the Valley.
4. Stem tubers. Examples: Irish potato and Jerusalem artichoke.
5. Fleshy roots. Examples: sweetpotato and dahlia.

Importance and Uses. It is appropriate to outline briefly several phases of interest related to the production and uses of bulbs and other modified structures considered in this chapter:

Propagation Stock. The kinds of plants included in the foregoing classification are planted widely by commercial growers and home owners. This, then, creates a need for planting stocks. Formerly most of the planting stock was grown in foreign countries, notably in Holland. In recent years restrictions have been placed on importations of foreign-grown stock to lessen the danger of introducing new disease and insect pests. The effect of these restrictions has been to stimulate domestic production of bulbs to supply the need for planting stock. Commercial producing areas have developed in Florida, Virginia, Michigan, California, Washington, and other states, and the domestic supply is now being produced largely in these areas.

Forcing. Certain kinds of bulbs are used widely for forcing to produce blossoms, usually at seasons when flowers are not plentiful. They are popular with commercial florists, who sell

Fig. 64. A naturalized planting of narcissus bulbs.

them as cut flowers or as potted plants. Bulbs are the most popular plants for forcing in the home. They may be grown in bowls of water, in sand, or in soil.

Naturalized Beds. Bulbs are popular flowers for naturalized beds, where they are allowed to grow largely undisturbed for a number of years. They are especially suitable for this purpose because they are hardy and are attacked by relatively few insect pests or diseases, they normally bloom at a time of the year when flowers are scarce, and they grow several years with a minimum of care.

Economic Use. Several of the plants of this class have economic value because of the edible parts which they produce and their importance as food plants. This is notably true of the Irish potato, sweetpotato, onion, banana, and asparagus.

Bulbs. Most bulbs are subterranean. Some, however, are borne above the ground.

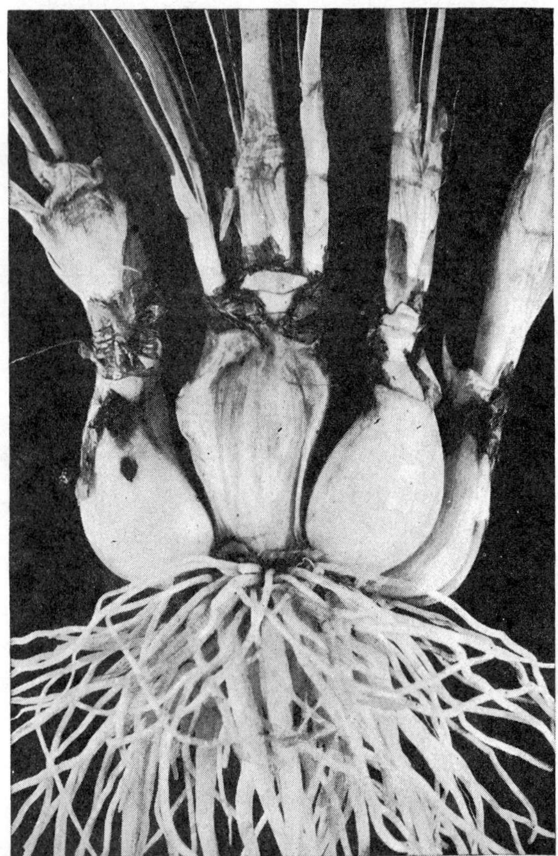

FIG. 65. Cluster of narcissus bulbs, showing mother bulb and two slabs on the right and one on the left.

Structure. A bulb is a modified stem in which the central axis is vertical and much shortened, perhaps to ½ inch. The internodes and nodes are not easily distinguishable; the central axis has a terminal growing point and axillary buds. This would be expected since it is a modified, vertically compressed stem.

A bulb is comparable in structure to an ordinary bud which has the embryonic parts to produce a stem, and also to a cabbagehead in which the central axis, nodes, internodes, terminal growing point, leaves, and axillary buds are clearly evident.

The fleshy modified leaves are closely appressed. In some bulbs the modified leaves are continuous around the axis, forming a series of layers. In cross section these layers appear as concentric

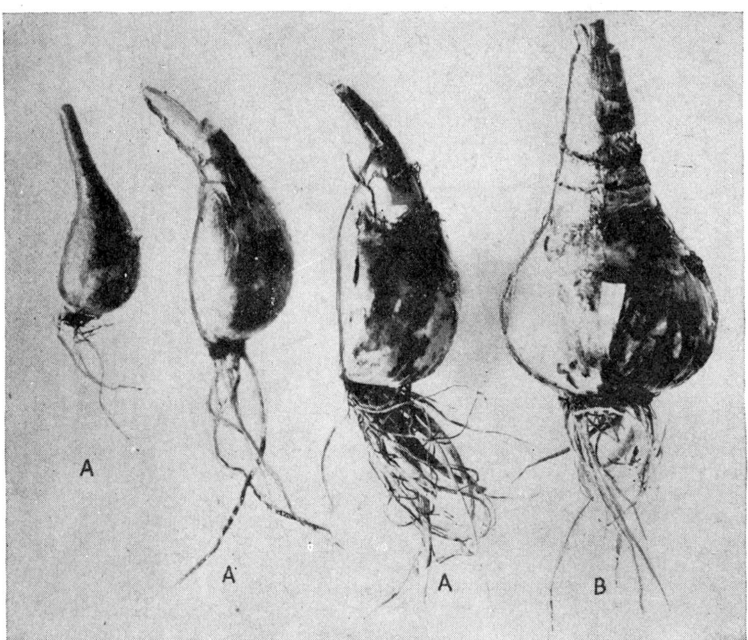

FIG. 66. Golden spur daffodil bulbs: *A*, slabs; *B*, round. (*Courtesy of Bureau of Plant Industry, U.S.D.A.*)

rings, as may be observed in the onion. Bulbs of this type are known as *layered* or *tunicate*.

In the other bulbs the scales are not continuous but are rather narrow and fleshy; they may be removed singly from the outer edges of the bulb. They are known as *scaly bulbs*, and lilies are the most important members of this group.

Growth Cycle. When bulbs are planted, the following growth processes are likely to occur: adventive roots develop from the base of the central axis; growth of the central stem at its terminal produces more leaves *in the interior of the bulb;* the bases of these leaves become additional layers or scales; under favorable conditions, a flower stalk is produced by the terminal growth and elongation of the central axis; axillary buds present in axils of modified leaves on the central axis may grow and produce other new bulbs, with all the characteristics of the mother bulb. These provide a means whereby bulbs may be increased in numbers.

The narcissus group of bulbs, which includes daffodils and jonquils, comprises a group of very useful and popular flowers. The

Fig. 67. A clump of tulip bulbs which has developed from one mother bulb. (*Courtesy of Bureau of Plant Industry, U.S.D.A.*)

normal cycle of reproduction of the narcissus requires a period of 3 years.

The mother bulbs, as they reach maximum size, develop buds in the axils of the layers. These buds, still attached to the central stem, continue to develop, forming daughter bulbs, or "slabs," which may be separated easily from the mother bulb at the end of the growing season. These, separated and replanted each year for 3 years, become successively larger until flower stalks are produced and the cycle of development is complete.

In certain kinds of layered bulbs, the mother bulb is depleted

Fig. 68. A large marketable tulip bulb, on the right, and three splits, on the left, which will produce bulbs of flowering size the next year. (*Courtesy of Bureau of Plant Industry, U.S.D.A.*)

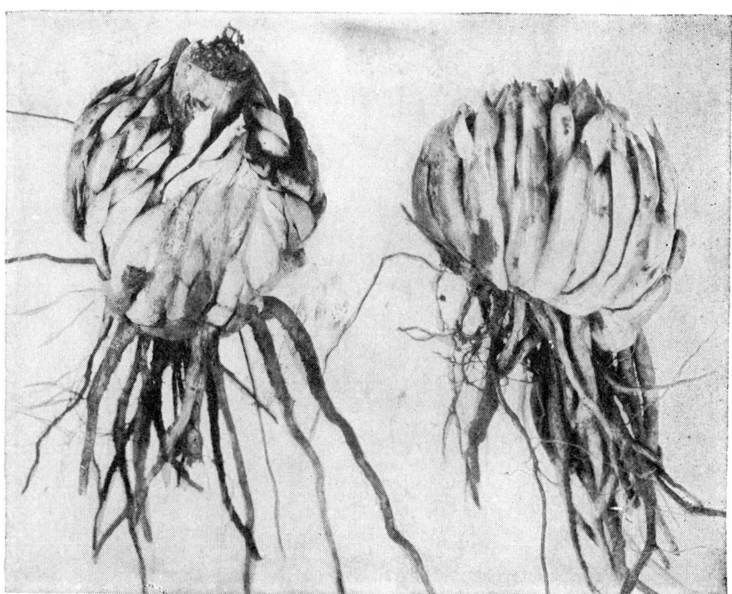

Fig. 69. Typical lily bulbs. Old flower stalk is shown in bulb on left. (*Courtesy of Bureau of Plant Industry, U.S.D.A.*)

each season of growth, and bulbs for further propagation are derived entirely from those that form from axillary buds. The tulip is an example of the bulb of this type. The formation of a large number of adventive bulblets can be stimulated in hyacinth bulbs by cutting into the basal portion of the mother bulb to remove the entire basal plate, or cutting across the base deep enough to extend through the growing point.

The lily is the most important of the scaly bulbs. It is an unusual plant because of the several different methods by which it can be successfully propagated. A large bulb will have from 75 to 100 *scales*. The scales, when detached from the mother bulb and planted under suitable conditions of temperature and moisture, will develop small bulblets on the inner, or concave, sides. They arise from adventitious buds. These are separated from the scales in due time, and when grown under suitable conditions, will develop into normal-size bulbs. The time required will be from 3 to 4 years. *Stems* of some species will produce a large number of new bulblets. They are pulled from the old bulbs and heeled in shortly after the flowers have opened. In a period of 35 to 40 days, small bulblets will have formed on the base of the stem. Their origin is largely from adventitious buds. These may be removed and planted singly, or the entire stem with the bulblets intact may be planted horizontally to provide increased growth of the small bulbs. *Cuttings* of the stems may also be made, with three or four leaves intact, or individual leaf cuttings may also be made, with heels or mallets of the stem. In either case, bulblets are formed from axillary buds. They are separated and grown until they reach flowering size as indicated above.

Aerial bulbils are formed by several species of lilies. They occur in the axils of the upper leaves and may be removed soon after the flowering period. The bulbils may be set in beds and allowed to grow for 2 years, by which time some of them will be producing flowers.

Division of the bulb occurs under natural conditions as a result of growth of axillary buds, and small increases may be obtained by digging the bulbs at intervals of 4 to 5 years for division. In commercial propagation, one of the other methods described will give more satisfactory results.

Seed are produced by almost all the lilies, and this method has the advantage of producing immense numbers of new plants.

It also is a means of producing new varieties. The growing of lilies from seed is a delicate undertaking. Seed of some species germinate poorly.

Corms. A structure very similar to bulbs is the corm, or "solid bulb."

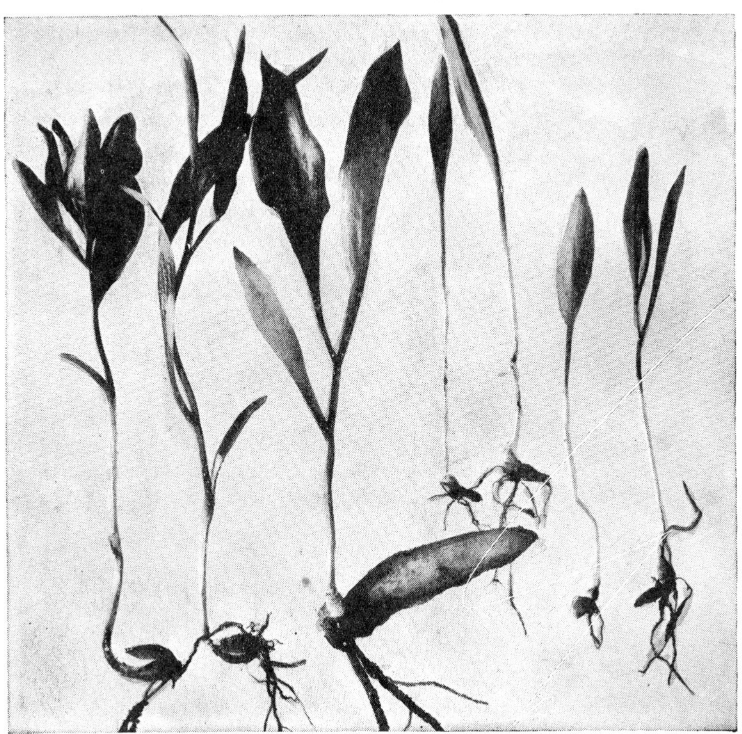

Fig. 70. Lilies grown from scales. Planted in July and photographed the following April. (*Courtesy of Bureau of Plant Industry, U.S.D.A.*)

Structure. A corm is a modified stem in which the central axis has become short and compact. The entire structure, when dormant, consists of the fleshy central axis. It differs from the true bulbs in that the dormant corm is solid, without layers or scales. In cross or vertical section, the structure appears as a mass of solid, undifferentiated parenchyma. The nodes and very short internodes of a corm are clearly evident. Apical buds and some axillary buds are present. At maturity, dried-leaf bases, arising from the various nodes, constitute the thin outer covering.

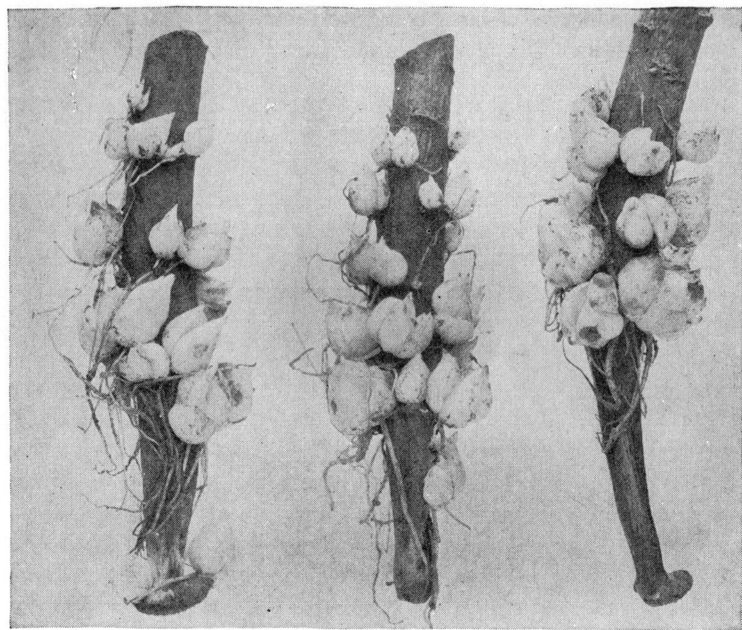

Fig. 71. New bulbs have developed on the base of the lily stems which were heeled in the field from July 15 to October 15. (*Courtesy of Bureau of Plant Industry, U.S.D.A.*)

Fig. 72. The bulbs in this cluster have been allowed to grow undisturbed for several years. Note that the bulbs are crowded, and also note the large number of flowering stalks, which have been cut off. (*Courtesy of Bureau of Plant Industry, U.S.D.A.*)

Growth Cycle. Gladiola and crocus, examples of plants that produce corms, are normally planted in the early springtime. They are also popularly planted in greenhouses at all seasons to produce flowers.

Shortly after the corms are planted, adventive roots develop from the center area of the lower surface. Usually one apical bud initiates growth and produces an aerial stem. These stems may reach a height of from 2 to 4 feet. They have leaves, nodes, and internodes, and usually produce an indeterminate inflorescence at the top with numerous sessile flowers being borne laterally. These aerial stems and flowers are the marketable parts of the growing plants.

In the growth of a corm and the development of the aerial stalk, the old corm becomes depleted. It is replaced by another one which develops immediately above it, forming the base of the aerial stalk. The aerial stalk withers at the end of the growing season and the large corm, when it matures, can be stored for future planting. In addition to the one large corm, a great number of smaller ones will develop adventively from the base of the large corm. Some are attached directly to the larger corm, others are borne on short rhizomelike structures. These small corms, or cormels, if planted, give rise to corms that become progressively larger and reach blooming size in 2 or 3 years, depending upon their original size.

Rhizomes. Some of our most interesting plants are propagated by rhizomes, the most common ones being monocotyledons. Interest in the structure and propagation of plants by rhizomes is derived from a desire to reproduce certain worthwhile species and to eradicate or control others that are undesirable.

Structure. A rhizome is the least modified of the subterranean structures. It is simply a stem growing in a horizontal direction slightly below the surface of the soil. It has the same general structure as the typical stem; nodes and internodes, axillary buds, and a terminal growing point are clearly evident on most, though not all, rhizomes; rudimentary or scaly leaves are present on some but lacking on others.

Growth Processes. Rhizomes develop from seedling plants by the growth of axillary buds on the base of the aerial stem. Once formed, the rhizome may continue its growth below ground by the continued elongation of the terminal growing point; or the terminal bud may grow to the surface and produce an aerial stem at any time. Branch rhizomes arise from axillary buds either on rhizomes or the bases of aerial stems. Thus a rhizome is capable of producing either aerial stems or new rhizomes from either the terminal growing point or from its many lateral buds.

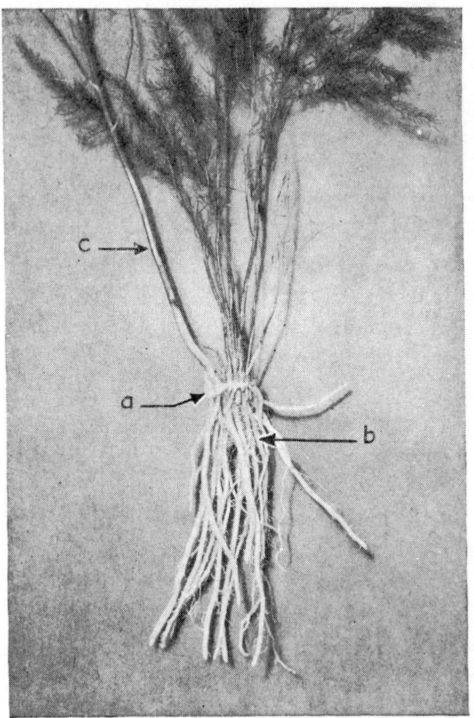

Fig. 73. Asparagus plant, showing rhizome (*a*), fleshy roots (*b*), and aerial stems (*c*).

Rhizomes produce adventive roots principally from the nodes. These form readily on even detached parts of rhizomes of most plants. The ready formation of roots, the abundant supply of stored food which rhizomes normally contain, and the readiness with which growth occurs to produce aerial stems and other rhizomes are reasons why new plants can be grown easily from rhizomes. These same factors are responsible for the difficulty encountered in eradicating undesirable types of plants that have rhizomes. Johnson grass (*Holcus halepensis*) is a troublesome pest in cultivated crops and is very difficult to eradicate because its rhizomes live through unfavorable seasons and because of the rapid growth of all parts of the plant during even a short favorable season.

Asparagus plants started from seed very soon develop rhizomes. These are highly modified in structure so that nodes

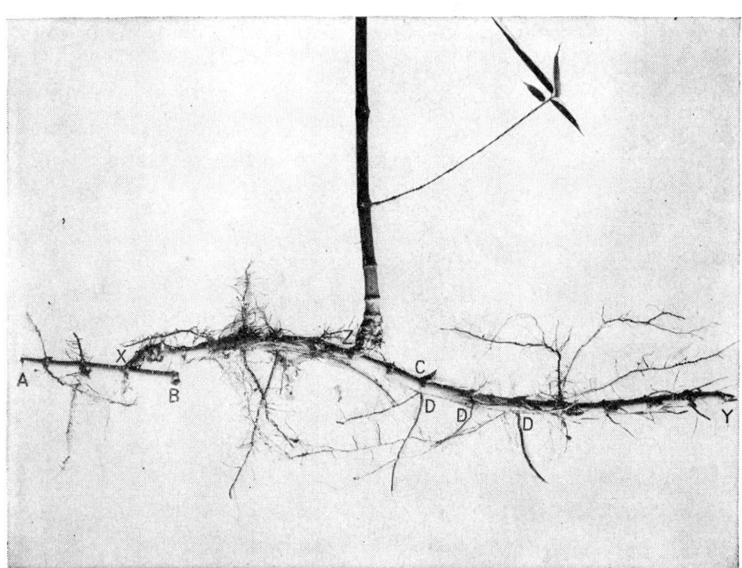

Fig. 74. Old rhizome of bamboo, *A* to *B*, produced new branch rhizome, *X* to *Y*, from lateral bud at *X*. Lateral bud at *Z* produced an aerial stem. Dormant lateral bud shown at *C*, and roots shown at nodes *D*.

and internodes are obscured. The rhizome grows very slowly, 1 to 2 inches per year. From the rhizomes are produced thick fleshy roots that contain reserve food materials, and they in turn give rise to fibrous branch roots that function as absorptive organs. Lateral buds are produced in clusters from the upper side of the rhizome, and at certain seasons these grow to produce the aerial stalks which, when young, constitute the edible portion of the plant. New plants grow readily from parts of the rhizome obtained by dividing it into segments.

Stem Tubers. The Irish potato is the best-known example of a plant that produces stem tubers.

Structure. A stem tuber is a shortened, thickened, underground stem. The "eyes" of the tuber are the modified axillary buds. The stem that bears the tuber is a thickened, underground, lateral stem, with nodes, internodes, and axillary buds. It is known as a *stolon.*

Growth Cycle. Tubers contain quantities of stored food available for supporting the initiation and growth of new stems and new roots. When the tuber is planted under favorable conditions, axillary buds start growth and produce aerial stems. Those near the apex of the tuber grow more readily than those near the base, a condition known as "apical dominance." Axillary buds below ground on the aerial stem, under proper conditions of temperature and darkness, initiate growth resulting in stolons. In its development, the stolon becomes thickened at the end and produces the tuber, which continues to enlarge until its growth is restricted by unfavorable growing conditions.

The first growth of a potato plant is supported by stored food in the tuber that is planted. Later, adventive roots develop from nodes below ground at the base of the aerial stem to support continued growth of the plant through the season.

Fleshy Roots. Sweetpotato, carrot, and beet are examples of plants characterized by fleshy roots. Roots of the former are *lateral,* whereas those of the later two are *fleshy taproots.*

Structure. Fleshy roots differ from stem tubers in that they do not have organized buds present on any part of them. As commonly harvested, beet and carrot fleshy roots are combined with the short stem portion of the plant. Fleshy roots contain an abundance of stored food to stimulate and support new growth and development of stems and roots.

Growth Cycle. When planted under favorable conditions of temperature, moisture, and oxygen, fleshy roots of the sweetpotato produce new aerial stems, commonly known as *slips.* The origin of these is from adventive buds, since there are no true buds on the fleshy root. These slips occur more freely on the basal end of the root, a condition known as *basal dominance.* The slips are typical stems of the species and have nodes, internodes, leaves above ground, rudimentary leaves below ground and axillary buds.

When a slip is detached from the parent root and planted properly under favorable conditions, it produces adventive roots from the lower portion of the slip, in contact with soil. These occur at nodes, at points lateral to the axillary bud. Vine cuttings are sometimes used for planting a field of sweetpotatoes, and root formation on these occurs in the same manner as when slips are used. This would be expected since the two have the same general structure.

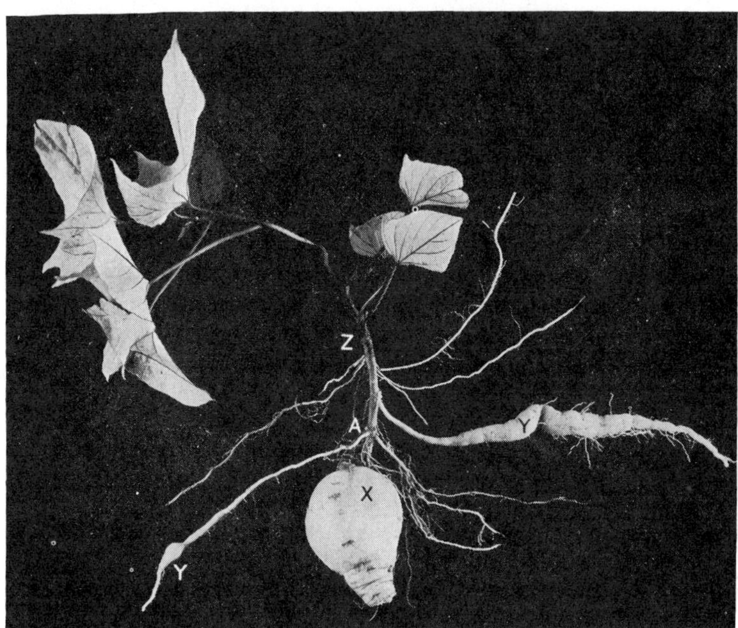

Fig. 75. Sweetpotato fleshy root, *X*, produced slip, *Z*, which developed roots at *A*. Thickening of roots at *Y* will produce new fleshy roots.

The adventive roots which form from the stems, or the branch roots that develop in regular order from them, become thickened and fleshy as they grow, to form the fleshy root which becomes the commercial sweetpotato used for propagation and for culinary purposes.

Rest Period. The rest period in bulbs and the other modified plant structures is the interval between the time of harvesting and the time the structure will resume growth. This period is definite and pronounced for several of the important kinds considered in this chapter.

The buds of a recently harvested Irish potato tuber will not ordinarily grow for some time, even under favorable conditions. Fully matured potatoes have a shorter rest period than those harvested prematurely.

This explains the difficulty often encountered in getting potatoes of the late spring crop to sprout when they are used as planting stock for a fall crop. Storage at from 82 to 86°F. for 2 months is effective in ending the rest period. Treatment with certain chemicals, such as ethylene and chlorohydrin, hastens the ending of the rest period. Table-stock potatoes should remain dormant, and hence, conditions that make the rest period shorter are avoided in storing them.

Gladiola bulbs will grow best when a rest period is allowed the corm before planting. Greenhouse operators obtain planting stock for fall planting from areas where the corms were harvested in the spring and find that they grow better than recently harvested corms. The rest period of the corms can be shortened or ended by storing them at a temperature of 35 to 45°F., by soaking the corms in a solution of ethylene and chlorohydrin for 1 day, or by subjecting the corms to vapors of the same chemical for 3 or 4 days.

The narcissus, hyacinth, tulip, and several other types of structures have rest periods. The sweetpotato and rhizomes of Johnsongrass have no such inhibiting influence. Sweetpotato often sprouts in the field before harvest when the soil is excessively moist; and it commonly sprouts in storage.

QUESTIONS

1. What are distinguishing features of true bulbs?
2. What are the reasons for interest in bulb growing?
3. What are the ways in which a bulb is similar to a dormant bud in structure? In what way does a bulb resemble a head of cabbage?
4. What is the manner by which new bulbs originate from old bulbs? How do new bulbs originate from lily scales?
5. Outline the ways in which the following are similar in structure and the way in which they are different: bulb, corm, rhizome, fleshy root, stem tuber, stolon.
6. Outline the manner in which each of these grow and reproduce the structure: layered bulb, corm, rhizome, stem tuber, fleshy root, rhizome.

SUGGESTED REFERENCES

Brown, T. A.: Flowering Bulb Culture in Florida, *Florida Agr. Expt. Sta. Bull.* 48, 1928.

Griffiths, David: Commercial Dutch Bulb Culture in the United States, *U.S. Dept. Agr. Bull.* 797, 1919.

———: The Production of Narcissus Bulbs, *U.S. Dept. Agr. Bull.* 1270, 1924.

———: The Madonna Lily, *U.S. Dept. Agr. Bull.* 1331, 1925.

Magie, Robert O, and W. G. Cowperthwaite: Commercial Gladiolus Production in Florida, *Florida Agr. Expt. Sta. Bull.* 535, 1954.

Nakasone, Henry Y.: Breaking the Dormancy of Gladiolus Corms in Hawaii, *Hawaii Agr. Expt. Sta. Circ.* 41, 1953.

Shippy, William B.: Factors Affecting Easter Lily Flower Production in Florida, *Florida Agr. Expt. Sta. Bull.* 312, 1937.

CHAPTER 11

Graftage

Graftage is the art of inserting a part of one plant into another plant in such a way that the two will unite and continue their growth. It differs from cuttage, layerage, and bulb propagation in that the plant part expected to produce the top of the new plant is deprived of its own root system and unites with another plant that supplies this part.

The art of graftage is not new. Contrary to popular opinion, it is no recent innovation in the arts of plant craft. Pliny, writing before the birth of Christ, recognized graftage as a horticultural practice, and it is known that it was practiced before his time. Columella, who died shortly after the birth of Christ, mentioned certain kinds of graftage, particularly the bark graft, cleft graft, and patch bud, which he said had been practiced by the ancients. It is a significant fact that at those early periods in agricultural history the unreliability of seeds and the importance of graftage were appreciated in the reproduction of varieties. At various times, including the present era, many methods known in ancient times have actually been rediscovered by workers who were not familiar with their previous use.

The field of graftage includes *scion grafting* and *bud grafting*, commonly referred to as *grafting* and *budding*. The two, however, are so different that a discussion of each will be reserved for separate chapters. Some of the operations and terms that are common in all types of graftage will be considered in the following sections.

Top-working. The series of operations whereby the top of a plant is replaced with a top of a different variety is known as *top-working*. In some cases a large part of the old top is cut away

148

and a new one started; in others the new top is started, after which the old one is cut away by degrees until the top consists largely of a different variety. Trees may be top-worked successfully by either budding or grafting or by a combination of the two. The process may be completed within one season or it may extend over several years, depending upon the size and conformation of trees to be top-worked. In reality, budding or grafting of small nursery trees is top-working; the term, however, is generally used with regard to changing the tops of larger trees.

Dehorning. The practice of cutting the main limbs and trunk of a tree back to stubs is known as *dehorning*. The extent to which trees can be safely cut back varies with the species; some can be cut back much more severely than others. In practice, trees are cut back so that the stubs that remain range in length from 1 to 4 feet and in diameter from 1 to 6 inches. When a tree is dehorned, the limbs should, if possible, be cut at points that will result in the new top having a symmetrical shape. It is not advisable to cut limbs at different heights so that the new growth of some will obstruct sunlight and create shade for others. In order to facilitate healing of the wound, a limb to be dehorned should be cut at a point where a side limb or a lateral bud occurs on the upper side. This virtually assures growth from very near the terminal part of

Fig. 76. An oriental pear tree top-worked with the Garber variety. Seven buds were inserted in the framework of a 3-year-old rootstock early in April, and the picture shows them as they began growth the following March. The points marked with an X indicate places where the T buds were inserted.

the stub, and this encourages overwalling of the wound. If no shoot grows within 1 inch of the end of the stub, it usually becomes advisable to recut it during the first year at a point where a lateral limb has developed in the meantime, preferably on the upper side of the stub.

The season for dehorning is during the dormant period, shortly before growth is resumed in the springtime. The practice has

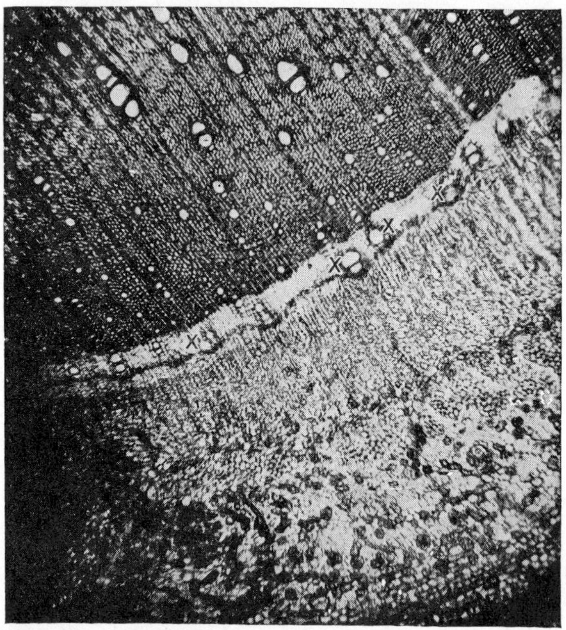

Fig. 77. Isolated vessels (greatly enlarged) in the cambium region shortly after growth starts in the early spring, a condition unfavorable to successful budding or grafting.

considerable application in renovation pruning and is used in many cases as a preliminary step in the top-working of large trees.

Forcing. Any treatment that encourages and hastens growth of a bud or graft is referred to as *forcing*. It is known that the terminal growing point creates hormones which restrain the growth of lateral buds below. Essentially, forcing consists of elimination of this influence and creating for the bud or graft a terminal position from a physiological standpoint. In practice,

FIG. 78. Peach tree of appropriate size for top-work.ng.

FIG. 79. Same tree as shown in Fig. 78, dehorned in late winter. New shoots which develop will be budded during the following summer.

it may be accomplished by (1) cutting the stock off above the bud, by (2) girdling above the bud, by (3) allowing the tying material to bind above the bud, and by (4) bending or breaking the stock above the bud without severing it. Buds or grafts that are exposed to sunlight are more easily forced into growth than those that are shaded.

FIG. 80. Pecan tree of suitable size to be top-worked.

The size of the stock and growth habits of a plant should determine the extent to which forcing is practiced. If forced too much, there is the likelihood of the graft or bud becoming top-heavy and breaking during a wind, rain, or ice storm. This is especially likely to happen if the stock is large and inflexible; the pressure of a force in such a case is applied at the point of union where vascular tissue connections may not be strong enough to withstand the strain. Terminal grafts, because of their position, seldom require forcing; their growth, however, can be influenced by the extent to which native shoots on the stock below the graft are cut back.

Stocks. The term *stock* has several different meanings when used in connection with propagation. *Stock plants* are those that are grown, frequently in a greenhouse, as a source of propagation material, such as cuttings, layers, and sometimes buds or grafts. *Nursery* stock means plants that are grown by nurserymen, usually for sale. *Lining-out* stock refers to plants that are of suit-able size to be replanted or lined out in the nursery row and allowed to grow for one or more seasons, during which time they become large enough for a designated use. A *rootstock* is the plant that supplies the root system, and in many cases a part of the trunk and framework, for a budded or grafted tree; root-stocks are commonly grown from seeds, cuttings, and, rarely, leaves. The term *stock*, as frequently used, is synony-mous with rootstock. The term *rootstock* is also used to describe underground stems, such as *rhizomes*, and this causes some confusion in in-

Fig. 81. Pecan tree dehorned to cause growth of new sprouts suitable for budding (see Fig. 80).

terpretation of horticultural literature. In propagation work, plants that are grown from seed are known as *seedlings*, until they are budded or grafted, after which they become *seedling* rootstocks. Those plants that are propagated by asexual methods, usually cut-tage and occasionally layerage from a single plant or variety, are known as *clonal* rootstocks after they have been successfully budded or grafted. When grafts are placed on unrooted cuttings, the latter are called *cutting stocks*. An *intermediate* stock is one that separates and connects a rootstock and a budded or grafted top, yet is different from both; it is also known as a *splice*. The term *body* stock is used to indicate that the trunk and framework of a tree are different from the budded or grafted top. The body stock may be the same as the root system, but in many cases it is of a different kind, as a consequence of double-working.

Matrix. The matrix is a place on the rootstock that is prepared for the insertion of a bud or graft.

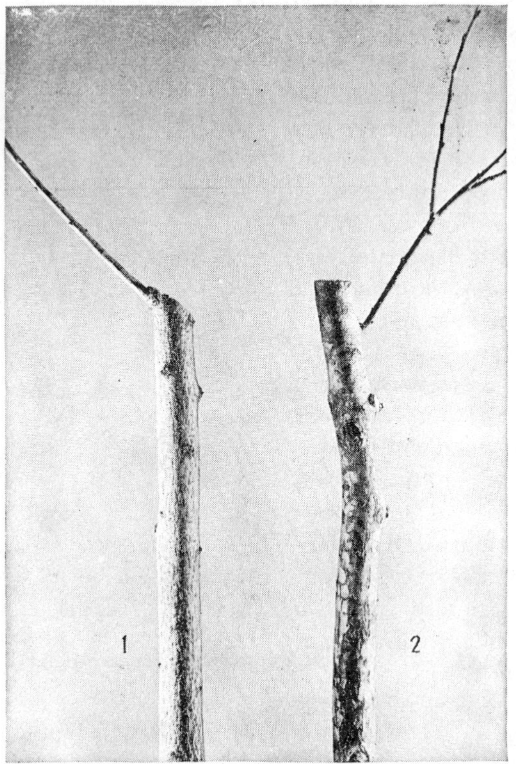

FIG. 82. Showing (1) proper and (2) improper way to cut limbs in dehorning.

Scions. The limbs that are cut from any plant to be used in graftage are known as scions; those which are to be used for grafting are known as *grafts,* or *graftwood;* and the ones that are to be used as a source of *buds* for budding are called *budwood.* Scions reproduce the kind of tree or plant from which they are taken and hence are obtained from the variety to be propagated.

Healthy parent plants should be selected in order to prevent the spreading of disease in propagation.

The success of budding and grafting by the different methods depends, among other things, upon the use of the appropriate kind or type of scions, and also upon methods of handling them from the time they are cut from the parent tree until they are finally used. The time that intervenes may be a few hours or several months.

Graftwood. Scions for grafting are usually obtained from one-year-old wood; sometimes older wood is used. They should be straight, smooth, have normal, plump buds, and few or no side branches. The size range for graftwood may vary considerably for different methods.

Scions for grafting should be thoroughly dormant at the time they are used. Dormant scions normally contain reserve stored food to provide energy for respiration, callus formation, and early growth of the scion. They should be secured before the plant from which they are to be taken shows any signs of growth. Those cut in midwinter tend to remain dormant longer after they are inserted into a stock than those that are cut at a later date; scions that make premature top growth before union is established usually wither and die within a few days. In practice it is customary to cut scions any time from midwinter until 2 or 3 weeks before the parent tree begins growth in the spring. These grafts are frequently used for grafting as they are secured from the parent tree, with no more than a day or two intervening.

Scions that are to be used relatively late in the grafting season may be cut and held in cold storage in order to keep them dormant. At a temperature of from 32 to 36°F. graftwood can be stored successfully for 4 months or longer, though there is seldom any occasion for storing it this long. Prior to storage, the scions should be packed in moist insulating material. Sphagnum moss and coarse sawdust are commonly used for this purpose; both are light and easy to handle, absorb moisture readily, and retain it well. The packing material should be moist, though not so wet as to cause poor aeration. Tests with storing scions show that from 3 to 5 pounds of water for each pound of sphagnum moss is sufficient. It is desirable to add the water and soak it uniformly into the insulating material before packing it around the scions. Alternate layers of the insulation and scions are placed in a box of convenient size for storage. It is important that the packing be pressed firmly into all corners and around the edges of the box. Heavy paper folded over the box will restrict evaporation and delay the time when additional moisture will be needed. If packed properly and held at a low

Fig. 83. Showing method of packing budwood to be placed in cold storage; another layer of moss is placed over the budwood.

temperature, scions remain thoroughly dormant and may be used directly out of storage throughout the springtime.

Scions should be properly labeled with the name of the variety at the time they are cut; otherwise different varieties are likely to become mixed. Failure to preserve the identity of scions is serious, because mixing of varieties is oftentimes not discovered until the trees come into bearing.

Budwood. Several different types or kinds of scions are used as sources of buds for budding. The classification of these is based largely on the age of the wood and its condition of growth or dormancy at the time of cutting or use.

Current-season Scions. Buds that are taken from limbs in their first season of growth are suitable for some methods of budding. When so used, they are known as *current-season buds.* The habits of growth of plants determine in a large measure how early in the growing season such buds may be used. On some plants they mature sufficiently for use within 6 weeks or 2 months

FIG. 84. (1) Dormant apple scions; (2) current-season budwood of rose before and after the leaves are removed.

after growth begins in the springtime; on others they do not mature until much later. In some areas it is customary to use peach buds of the current season's growth as early as June; in the North they are not sufficiently mature until a considerably later period.

Current-season budwood may be "ripened" by cutting, at a point ½ or ¾ inch out from the base, the leaf petiole subtending each bud. This is done before the scions are cut from the parent tree; it causes the petiole to fall off within 6 or 8 days, and a corky covering forms over the petiole scar. If the leaf petioles are cut too far in advance of the time for cutting the scions, buds in corresponding axils are likely to force into growth and not be suitable for budding purposes. In ripening budwood the petioles should be cut from only the basal part of each shoot, so that perhaps only two-thirds of the leaves are removed. The buds near the terminal are usually too immature for use, and there is no object in removing the leaves from that portion.

All leaves should be removed from current-season scions as soon as they are cut from the tree in order to restrict transpiration. Buds from them should be used as soon as possible after they are obtained. If they cannot be used immediately, they may be kept for a week or longer, packed in moist material, and stored at a temperature from 34 to 38°F.

Previous-season Scions. For early-spring budding it is necessary to use budwood that grew during a previous season. Limbs that have made normally vigorous growth during the preceding season are preferred. The basal and midportions of one-year-old wood furnish the best buds; those buds on the angular, small, or immature wood of the terminal portion are not usually satisfactory. Buds may be used from two- or three-year-old wood of some plants, but they are seldom entirely satisfactory. There are two different methods of using previous-season wood:

1. *Fresh scions* are cut from previous-season growth of the parent tree as needed throughout the season for budding. Budwood of some plants remains in acceptable condition for use throughout the season; on other plants the best buds are forced into growth in the early springtime, and buds that are satisfactory for use after that time are scarce.

2. *Storage budwood* is used extensively in the propagation of certain plants. Two general practices are followed in using it. According to one of these, it is cut during the dormant season and packed and stored under conditions similar to those prescribed for graftwood, and at the same temperature. For certain methods of budding it may be used directly out of cold storage in a dormant condition. If, on the contrary, it is to be used in one of the methods which requires that the bark separate from the wood, the budwood must be subjected to conditions upon re-

moval from storage that will cause the cambium layer to become active. This treatment, known as *seasoning,* is accomplished by providing ample moisture to prevent drying out and a temperature of from 78 to 85°F. It is considered that budwood is seasoned when the bark and buds can be peeled readily from the wood. The number of days required for budwood to become seasoned varies commonly from 3 to 10 days. Seasoned budwood may be used immediately; or it may be returned to cold storage and held for as long as 1 month. It should be noted here that dormant budwood may be kept for several months but seasoned budwood for a limited period only.

By another practice previous-season budwood is cut, packed, and stored as soon as the bark will slip but before the buds have made any perceptible growth. Budwood of this kind will keep satisfactorily for about 1 month at a temperature of 32 to 36°F. The cambium layer remains active in cold storage and the wood can be used as soon as it is removed.

The Bud or Graft Union, and Compatibility. The immediate objective of graftage is to secure the union of the scion and rootstock. After this the growth of the budded or grafted top depends upon the compatibility of the scion and stock.

Union. The union of the scion and stock takes place as a result of the formation and commingling of callus on the two components. The callus is produced by the cambium layer of plants as a spongy mass of unorganized parenchyma cells. The manner in which the callus forms and its general direction of growth determine whether union is accomplished by regeneration or overwalling, or a combination of the two.

When the scion is placed in the matrix on the stock, a definite effort is made to have the two cambium layers match at least on one side. If, as a result of poor technique, the scion and stock fit poorly, a greater amount of callus will be necessary, a longer time will be required, and the chances of ultimate union are less than if a better fit were obtained. A firm pressure is always necessary to produce a graft union, because the callus formed by the stock and that formed by the scion tend to spread or separate the two component parts. For this reason they must be held in place by tying if necessary, until a union is formed. Many bark grafts especially are lost because the strings are removed too soon. Union of a graft and stock takes place only

FIG. 85. The large limb shown at X is live oak (*Quercus virginiana*) top-grafted on post oak (*Q. minor*). It grew vigorously when first grafted, but became progressively incompatible, as indicated by thin foliage, during a period of 17 years.

in tissues that form after the graft is made; the original woody portion of a scion never unties with a stock. The proper placing of a scion affects not only the probability of its growing but also the strength of the resulting union.

Several factors and conditions influence callus formation

Some *species* form a large amount of callus and others smaller amounts. Some form callus quickly and these are likely to unite if properly grafted, while those that form callus slowly are less likely to unite. Plants used for rootstock or scion that are well supplied with *stored food* form callus more readily than those that are poorly supplied. This would be expected since stored food is the source of energy from which callus is produced.

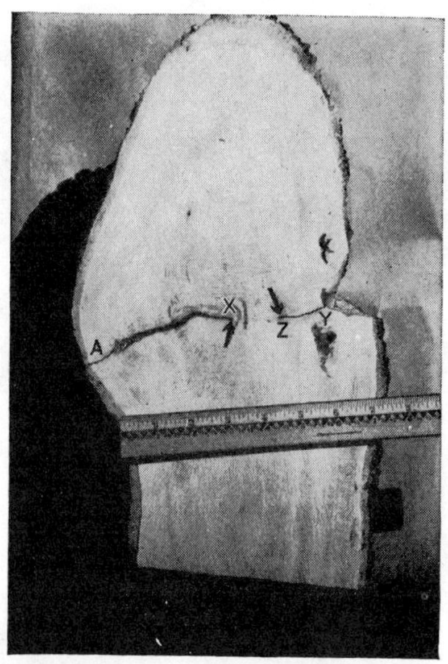

Fig. 86. Longitudinal section of the live oak graft shown in Fig. 85. The tissues between the arrows that developed during the early growth of the graft are continuous, but there is no union beyond arrows on either side. This explains why the graft made weak growth and had thin foliage during the latter years before it died.

Scions for grafting are cut while dormant because in that condition they have the greatest supply of stored food. Callus formation is influenced by *temperature*, 70 to 75°F. being considered favorable for such plants as the apple and grape. This explains why outdoor grafting or budding is seldom practiced in midwinter. The continued growth of callus tissue depends upon a high *humidity*. Various treatments are used to provide this for the regions where callusing takes place. The buds or grafts may be tied with materials that will keep them moist, such as rubber budding strips or waxed tape, or they may be waxed; whip grafts are packed in moist insulation material, or moist soil is packed around them in the nursery; some grafts are covered with plastic film bags to retain moisture; some buds, such as the T bud, are inserted and tied in a way that prevents drying of the cambiums.

Plants that produce gums, resins, latex, and tannin are usually difficult to bud or graft because these inhibitors tend to prevent

Fig. 87. Graft union showing overgrowth of rootstock.

proper contact between the callus tissues from the stock and scion.

Monocotyledonous plants are never grafted on a commercial basis because they do not have continuous cambium and form little callus tissue.

Compatibility. The term *stion* is used to designate the relationship between a stock and scion; *stionic effects* are the reciprocal influences of stock and scion, and they determine the *compatibility* of the two. The fact that a stock and scion will unite is no assurance that the union will be strong or enduring.

Oftentimes the response of a given top on a root system is not predictable on the basis of botanical relationship. Some plants will unite readily but very *quickly* become incompatible. This is true of the pear on quince and of the apple on red haw. There are others which unite and make normal growth for a prolonged period, but which ultimately fail as a result of *delayed* incompatibility. The immediate cause of failure may be due to either the rootstock or the scion top. Failure of either is quite soon reflected in the functioning of the other. The basis of the suc-

Fig. 88. Bruce plum budded on American plum rootstock, showing marked overgrowth of top.

cess are failure of any stionic union is either anatomical or physiological in the final analysis. *Anatomical* differences are reflected in the characteristic overgrowths of the stocks in some cases, and scions in others; and also in the alternating unions and "breaks" that occur in grafts between certain plants in the cambial region as the point where the two meet. *Physiological incompatibility* is due to inability of the stock or scion adequately to supply the other component with the necessary amount or quality of materials for normal functioning. The so-called black-end disease of some pears on certain Oriental rootstocks is apparently an example of this. Whatever the relationship with reference to stionic compatibility may be, the effect is to influence the vigor of the plant. Vigor, in turn, influences indirectly the quality and size of fruit; yields; earliness of bearing; earliness of maturity; hardiness to heat, cold and

FIG. 89. Yellow Transparent apple grafted on an incompatible rootstock, the native mayhaw (*Cretagus* sp.), produced 16 apples during the third season. (*Courtesy of A. W. Orr, Livingston, Tex.*)

drought; and stature, some being dwarfed by a certain rootstock and invigorated by another.

As a general rule a horticultural variety may be budded or grafted upon any other variety of the same species and the unions will be strong and enduring. The Elberta peach will unite with the J. H. Hale peach and the Delicious apple with the Jonathan, for example.

It **is** generally true that species of the same genus may be cross-grafted, or budded, one upon the other. Peach will grow on plum, almond, cherry, or apricot, for example, forming good unions with some and poor unions with others; pear will unite with the apple; and California privet and Amur privet are used as rootstocks for waxleaf ligustrum. All these are examples of

Fig. 90. Kieffer pear on Japanese pear rootstock. Budded shoot on the left center had begun growth in the spring, while the stock was apparently still dormant.

plants of species that will unite with other species of the same genus, some forming successful and others unsuccessful combinations.

Plants that belong to different genera of the same family cannot be intergrafted with any certainty of success. There are examples of some that will unite, and many others that apparently will not. The apple unites with the hawthorn, the chestnut with white oak. In general such wide crosses do not form satisfactory combinations. Chestnut on oak seldom lives longer than one or two seasons. Apple tops may live for several years on hawthorn rootstocks, but they seldom grow or fruit in a normal manner. It is not possible to graft a stone fruit, as the peach, onto a pome fruit, as the apple or pear, by methods ordinarily used in propagation of these plants.

There are few, if any, records of plants belonging to different families that will unite when budded or grafted.

Double-working. Double-working is a practice in which two successive budding or grafting operations are performed on the same plant. After one scion has been placed on a stock, a second scion is grafted or budded into the first one. This procedure results in a tree composed of three different kinds or varieties of wood; the first scion inserted becomes a splice or intermediate stock between the root system of the original stock and the new top of the double-worked tree. The first scion is generally allowed to grow a year before it in turn is top-worked to another variety. However, buds of the desired variety may be set into the limb that is to serve as the splice before it is severed from the parent plant. The bud remains dormant until the splice is grafted onto the original stock, when it begins growth, producing a double-worked tree. The two grafting operations may be performed at the same time with some plants, though considerable skill is required to ensure successful unions. There are several applications of this method, especially in the production of fruit trees.

1. Uncongenial stocks and scions may be united by means of a splice that is congenial to both. Certain pear varieties, such as Bosc, Bartlett, and Winter Nelis, do not form good unions with quince, and if it is desirable to grow these varieties on quince rootstocks, they are double-worked by using a splice of Hardy, Angouleme, or some other variety that does make good union with the quince stock and the pear top.

2. Resistance to specific troubles may also be secured by double-working. Certain varieties of apple may be used as a splice to give resistance to collar rot. Freezing injury to the crown of the tree may be prevented or lessened by double-working to provide a cold-tolerant splice at the ground line.

3. Top-working of orchard trees that have been grafted origi-

nally results incidentally in double-working. In such cases the intermediate stock or stem continues to exert a certain influence on the root system of the double-worked tree and may also affect the new top as well.

The performance of a grafted or budded tree is determined by the reciprocal influences of both the stock and the top. The possibility of uncongeniality is increased where there are two graft unions instead of one, and three varieties of wood instead of two. Varieties should not be double-worked without regard to the intercongeniality of all three components.

Graft Hybrids or Chimeras. The term *graft hybrid* is a very misleading one; it fosters the rather common conception that hybrids are the normal result of grafting. On the contrary, one of the primary objects of graftage is to maintain a variety true to type.

A *graft hybrid* is a stem, branch, or plant originating from an adventitious bud at the graft union. Not all such buds produce graft hybrids—only those do so which contain tissue of both stock and scion. The resulting shoot and its fruit will be hybrids only from the standpoint of morphology and not from the genetic viewpoint. The graft hybrid has been designated by some writers as a special form of graft symbiosis.

Chimera is a term commonly applied to graft hybrids and to similar phenomena occurring naturally. In one case where the two masses of tissue meet, the plant and fruit will consist of sectors of tissue from both stock and scion; this type is known as a *sectorial* chimera. In another case, the tissues of stock or scion overgrow those of the other, to form a *periclinal chimera*. A third type, the *hyperchimera*, is produced by a vegetative cone that is a mosaic of unlike cells.

Objects of Graftage. Graftage may be employed to increase the usefulness of plants in a number of different ways.

1. It is a way of preserving and perpetuating some varieties that cannot be reproduced easily by other vegetative methods. When a relatively simple method like cuttage is not effective, graftage is often used successfully. Standard varieties of peaches, apples, walnuts, pecans, and many other fruits are not easily propagated by asexual methods other than graftage.

2. Graftage makes it possible to change trees of poor varieties to varieties that are considered to be more desirable. A tree

might be undesirable by being nonproductive, a shy bearer, or an irregular bearer; it might have undesirable habits of growth, an untimely blossoming period, or an untimely ripening period of fruit. Trees of considerable size when top-worked will resume production of fruit and produce more heavily in a shorter period of time than nursery trees.

3. Adaptation to unfavorable environment may be accomplished by graftage. In many cases, stocks may be found that are especially resistant to some soil conditions, as heavy soil, poor drainage, acidity, or alkalinity; to some insect or similar pest living in the soil, as grape rootlouse, woolly aphis, or nematode; or to some particular disease, as foot rot or gummosis of citrus.

Rootstocks vary in soil adaptation, tolerance of insects and diseases, and in other respects; and they should be selected for a variety with the same careful consideration given cultural practices.

4. Graftage is a means of hastening new varieties into bearing. Scions secured from young trees and caused to grow on mature trees reach bearing stage sooner than if left on their own roots. Several new varieties may be tested out on one large tree and thus space be conserved. This method may be used to advantage in fruit-breeding work.

5. The development of seed and fruit by dioecious plants may be encouraged by grafting or budding in scions of the lacking sex. Varieties that are incompatible or dichogamous may be rendered more fruitful by grafting in scions of varieties that are known to be good pollinators.

6. Graftage is used to encourage healing of tree wounds caused by implements, disease, injurious temperature, rabbits, mice, or other rodents.

7. The novelty of growing several different kinds of fruit or flowers on one plant is made possible by graftage. In response to popular interest, nurseries now offer single trees with plum, peach, and apricot; different varieties of apple and pear; various colors of roses; and many other similar combinations.

QUESTIONS

1. How long has graftage been practiced?

2. What is top-working? What determines the length of time required to top-work a tree?

3. What are the objects of dehorning? What rules should be observed in dehorning a tree?

4. How is forcing accomplished?

5. What is the meaning of stock plant? Nursery stock? Lining-out stock? Rootstock? Seedling rootstock? Clonal rootstock? Matrix?

6. Tell how to select, prepare, and store graftwood.

7. What are the different classes or types of budwood? How long can each be stored? What is seasoning?

8. Explain how union takes place between a stock and a scion. What factors influence the formation of callus?

9. What factors determine whether plants can be successfully grafted?

10. In what ways does the rootstock influence the grafted top of a tree? What are the causes of unsuccessful graft unions?

11. What are the various objects of graftage?

SUGGESTED REFERENCES

Armstrong, W. D., and F. R. Brison: Delayed Incompatibility of a Live Oak–Post Oak Graft Union, *Proc. Am. Soc. Hort. Sci.*, **53:** 543–547, 1949.

Bitters, Q. P., and E. R. Parker: Quick Decline of Citrus as Influenced by Top-root Relationships, *Calif. Agr. Expt. Sta. Bull.* 733, 1953.

Bradford, F. C., and B. G. Sitton: Defective Graft Unions in the Apple and the Pear, *Mich. Agr., Expt. Sta. Tech. Bull.* 99, 1929.

Brison, Fred R.: The Storage and Seasoning of Pecan Budwood, *Texas Agr. Expt. Sta. Bull.* 478, 1933.

Cummings, M. B., E. W. Jenkins, and R. G. Dunning: Rootstock Effects with Cherries, *Vermont Agr. Expt. Sta. Bull.* 352, 1933.

McClintock, J. A.: A Study of Uncongeniality between Peaches as Scions and the Marianna Plum as a Stock, *J. Agr. Research,* **77:** 253–260, 1948.

Reynolds, Howard, and J. E. Vaile: Effects of Rootstock upon Composition and Quality of Fruit of Concord, Campbell Early, and Moore Early Grapes, *Arkansas Agr. Expt. Sta. Bull.* 421, 1942.

CHAPTER 12

Methods of Grafting

Grafting is the operation of inserting a portion of a limb of one plant into another in such a way that the two unite. The methods of grafting differ with respect to (1) the season at which the operation is done, (2) the technique involved in fitting the scion and stock together, and (3) the adaptation to different kinds and sizes of plants.

Cleft Graft. The cleft graft is one of the oldest methods of grafting. It is also one of the simplest forms and may be used with good success by amateurs on trees that are easily propagated. The principal use of the cleft graft is found in top-working large trees where limbs from 1 to 3 or 4 inches in diameter are to be grafted. Cleft grafting should be done during the latter part of the dormant season just before active growth starts in the springtime.

With a sharp saw the limbs are cut off squarely at a place where they are straight and free from crooks and side branches. As far as possible limbs should be cut off at such places so that when grafts are inserted they will develop into a symmetrical top of good conformation. The end of the stub is then split with the cleft iron and is ready for the insertion of the scion. Scions are prepared by cutting the end in a wedge shape, with one edge of the wedge slightly thicker than the other. It is customary to make the cuts so as to leave a vigorous bud just above the top of the wedge and on the same side as the thick edge of the wedge. The cuts forming the sides of the wedge should be long, to ensure gradual rather than abrupt tapering toward the apex of the wedge. They are best made with a single full stroke of a sharp knife, and each cut should be a perfect plane. When prepared in this way, scions will fit into the cleft matrix better than if cut in a haphazard way or if the wedge tapers too abruptly. Two or three good buds on the scion above the crown of the stock are sufficient; the cut on the upper end of the scion should be made ⅛ inch above a bud. One scion is enough for each stock less than 1 inch in diameter, but it is a

Fig. 91. *Technique of cleft grafting*

common practice to insert two scions in larger stocks—one in each side of the cleft. On sloping branches the cleft should be made horizontally, so that the scions are lateral to each other and not one above the other. In adjusting a scion, the cambium layer should be placed in contact with that of the stock at as many points as possible. The thick edge of the wedge of the scion should be toward the outside, so that the pressure exerted by the two halves of the stock will be at points where the cambium layers coincide. If the scions are not held securely in place, the pressure of the two halves of the limb should be supplemented with strong cord. The end of the stub and the entire length of the split are covered with grafting wax.

In the event two scions grow in one stock, the weaker one is kept headed back for a year or two and is finally cut off entirely even with the crown of the stock. In the meantime it will assist in the formation of new tissue over the end of the stub. Two scions are likely to form a weak, narrow V crotch if both are allowed to grow permanently in one stock.

Whip Graft. Whip grafting has long been a popular method of propagation and is the means whereby many different kinds of plants are reproduced. The method is used on stocks that are relatively small; those larger than ¾ inch in diameter cannot be conveniently whip-grafted because of the difficulty involved in making the cuts properly.

The top of the stock is cut off with a diagonal cut that should be about 1¼ inches long on ½-inch stock. This cut should be proportionately longer on larger stock and shorter on smaller stock. The scion, having been cut previously to a length of 5 or 6 inches, is cut across the lower end in a similar way to the stock. The cut on the stock should be made upward, and the cut on the scion should be made downward. Each of these surfaces should be smooth and as nearly a plane as possible; uneven or wavy surfaces prevent proper contact between the two cambium layers. A second cut is required on both stock and scion. It should begin one-third of the distance from the apex of the cut to the base and extend toward the base. This cut, which forms a tongue, should be slightly across the grain of the stock, its course being toward the base of the original slanting cut. When prepared in this way, the stock and scion are snugly fitted together, the two tongues interlocking. Extreme care should be exercised to have the cambiums of the two in contact with each other on one side. If the scion and stock happen to be of the same size, it is possible to have the cambiums coincide on both sides, in which case the chances for a successful union are increased. Little trouble, however, is experienced in obtaining a union if the cambium layers meet on one side only.

It is not customary to apply wax to whip grafts. They are

commonly wrapped securely with waxed string. Cotton twine may be used, but it requires tying whereas the end of waxed string is held in place by the wax. Waxed tape, masking tape, and nursery tape are becoming increasingly popular as wrapping materials for whip grafts because they discourage the formation of callus knots.

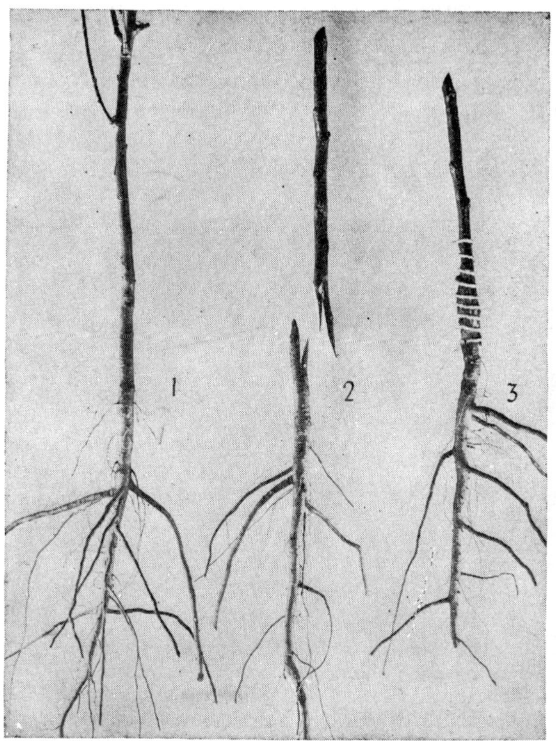

Fig. 92. Detail of whip graft. (1) Stock, (2) stock and scion cut, and (3) graft completed and tied.

Whip grafting may be done in the nursery, or the stocks may be dug and grafted indoors. This latter practice is known as *bench grafting*. Such grafts are usually packed in moist insulating material, stored in a room where the temperature is from 75 to 80°F. for 1 week or 10 days for callusing, and then planted in the nursery. If planting is delayed, the grafts can be held in cold storage until needed. In either nursery grafting or bench grafting it is customary to place the graft on the portion of the stock just below the ground line, and in replanting bench grafts the union should be planted slightly below the ground level.

Bench grafting may be done successfully at any time during the dormant season if a favorable temperature for callusing is provided. Nursery grafting should be delayed until the near approach of the growing season, because the temperature in the nursery is not likely to be favorable for callusing during the winter.

Bark Graft. The bark graft is unique in that the period during which it is done is after the bark begins to slip in the spring

Fig. 93. Cross section showing union of bark graft after one season of growth. Note that it formed union on one side only.

rather than during the dormant season. Scions used in bark grafting, however, should be thoroughly dormant. Stocks for bark grafting range from ½ to 4 inches in diameter. Scions may be from ⅓ to ¾ inch in diameter, depending on the size of stock on which they are to be used. With reference to position on the stock, bark grafts may be *terminal* or *lateral*, the former being used more commonly.

For a terminal bark graft, the stock is sawed off squarely, at a point where the bark is smooth and free from knots. The bark is split downward from the crown about 2 inches. If the bark is thick and rough, it is pared down so as to render it flexible on each side of the downward cut. On the lower end of the scion a sloping, downward cut is made, the surface of which should be 2 or 3 inches long, depending on the size of the scion. On the side of the scion opposite the lower portion of the cut just described, two additional cuts are made. They are each about 1 inch in length and are designed to expose additional cambium cells which make contact with similar cells of the stock and increase the chances of a union. These cuts intersect the first cut and result in a distinctly pointed scion, which may be readily inserted. Scions are usually 4 or 5 inches long and should contain two or three good buds. As soon as the scion is prepared it is inserted beneath the split bark of the stock. The long cut is placed in contact with the xylem or wood. The scion should be pressed down so as to allow only a small part of the cut to be exposed above the crown. Two or three scions may be placed in large stocks; thus the chance of growth is increased. In the event that more than one grows, only the most desirable one is allowed to remain permanently.

Tying with durable string and the application of wax complete the bark graft. Ordinary cotton twine may be used in tying on small stock, and stronger cord for larger stock. The tying material may be left on as long as it does not girdle, in order to support growing scions. Some propagators make a practice of placing twigs 2 or 3 inches long and the size of a pencil on each side of the inserted graft, and outside the flaps of bark which extend over the scion. In tying, the string is pulled over these two twigs. The effect is to press the bark of the stock in closer contact with the scion and decrease the air space on each side of the scion. The practice is especially recommended for large stocks, the bark of which has a tendency to bulge away from the stock when grafts are inserted beneath the bark. Small nails instead of twine may be used to hold bark grafts securely in place. Cigar-box nails are about the right size, and two are sufficient for each scion.

The essential difference between a terminal and a lateral bark graft is in position. There is little difference in the technique involved. A cut is made on the stock at a right angle to the main axis of the stem to a depth of $\frac{1}{2}$ to $\frac{3}{4}$ inch at a point where the bark is smooth and will permit the easy insertion of the graft. With a wood chisel or large knife the stock is notched above so as to intersect this cut. This notch permits easier insertion of the scion and better contact of it with the stock. The bark is split downward from the first cut made and a scion inserted. Details of preparing the scion and its insertion, tying, and waxing are the same for lateral grafts as for terminal. Foliage is left on the stock above lateral grafts, and this is thought to enable them to make better unions than terminal grafts. The rate of growth of lateral bark grafts can be regulated by the extent to which the native foliage beyond the graft is cut off; advantage is taken of this in preventing shoots from growing too vigorously and becoming top-heavy. The stock extending above the point of insertion can also be used later as a brace for the growing graft.

Inlay Graft. The inlay graft is similar to the bark graft in many respects. The size of stock on which it is used, season, and

Fig. 94. Technique of inlay grafting. Note that this one is to be made secure to the stock by nailing, instead of tying, after which it will be waxed.

graftwood are the same; but the graft itself is made differently. One large sloping cut is made opposite a good bud on the scion. This cut is to fit next to the xylem of the stock. On the side opposite the lower part of this first cut a second one that exposes additional cambium is made. The long cut surface of the scion is then placed on the stock and a strip of bark the width of the lower part of the scion is cut on each side. This flap is then peeled down and the scion inserted so that its long cut surface will be next to the xylem of the stock. The top part of the flap is cut off and the lower part is pressed over the outside cut surface of the scion. It is then tied and waxed. It is thought that inlay grafts make a stronger union than bark grafts. A double inlay graft is commonly used for bridge grafting which is described later.

Other Methods. The *veneer* graft is used chiefly in propagating (1) coniferous evergreen plants that are difficult to graft and (2) herbaceous greenhouse plants. A long sloping cut is made in the top or side of the stock, and in it the scion, which has been cut wedge-shaped on

Fig. 95. Inlay graft, used to top-work pear trees in April, had made growth shown here by July, several buds being allowed to develop from the scion.

the lower end, is inserted. When side cuts are made, the top of the stock remains intact until union is established, after which it is removed.

Cutting grafts are made by grafting a scion onto an unrooted cutting. Such grafts are usually stored as soon as they are made, under favorable temperature and moisture conditions, until the scion and cutting stock unite by callusing. They are then ready for field planting. If necessary they may be held for a few days or weeks properly insulated in cold storage before planting. Cutting grafts are sometimes planted in the field as soon as the scion and unrooted stock are joined together, in which event the two are expected to unite by callusing during the same period that the cutting stock is forming roots.

The *bottle* graft is an old method in which a long scion is grafted by the approach, or a similar method, to a stock plant growing in a pot, or to a branch where some sort of a support can be erected. The lower end of the scion is placed in a bottle partly filled with water, so that the scion may remain fresh and turgid until union occurs.

Root grafting refers to a grafting operation whereby a scion is grafted onto a root, used as a stock. In most cases whole root systems of small plants, except for slight pruning, are used; the scion is inserted in the upper part of the taproot, resulting in *whole-root* grafts. If one root is cut into two or more pieces and each is used as a stock, the grafts are called *piece-root* grafts.

A *nurse-root* graft is made by grafting a long scion onto a relatively short root piece. It is usually done indoors, and when the callused graft is planted in the field, the graft union is set deep, so that most of the scion is covered, all except the tip. The nurse root supports the initial growth of the scion. In the meantime roots are expected to form on the lower portion of the scion, in which case the plant becomes own-rooted, and the nurse root may or may not be cut off at a later date. When dwarfing species are used as rootstocks for nurse-root grafts, it is seldom necessary to remove the rootstock after the scion develops roots. Other means of inducing rooting of the scion involve the use of an inverted nurse root and girdling with string or wire at the point of union. A nurse root that is susceptible to the attack of a disease or insect pest may sustain a scion temporarily; if the scion is resistant to the trouble, roots that are formed on it will gradually replace those of the susceptible rootstock. Plants that root poorly from the scion may root satisfactorily from *scion shoots* that are caused to grow from below the soil surface by deep planting of the graft. Some plants are valuable for rootstocks because of having a good root system. If seeds do not reproduce a uniform rootstock of such plants and if they cannot be grown successfully from cuttings or layers, the nurse-root graft is a possible method whereby they can be produced on their own roots.

Aftercare of Grafts. Grafts in small stock usually require some training to cause them to grow into trees of the desired shape or form. Frequently two or more buds on a graft grow, with the result that a tree of undesirable shape is produced. The weaker growing bud should be cut off and all growth forced into one bud at first. After it has formed a single standard of desired height, lateral branches may be allowed to develop. Some plants are more valuable if they are branched, and these are trained accordingly. In order to ensure straight, upright growth of grafts, some propagators follow a practice of staking the trees with a lath.

Grafts in large stocks usually make a very vigorous growth during the first one or two seasons. They are likely to become top-heavy and to be blown down or broken off by high wind. This often happens, even where the stock and scion have made a good union, but it is more likely to occur if the union is poor or defective. Large stocks are rather inflexible, and the pressure exerted by wind against a graft is greatest at the point where the stock and scion meet.

Danger from loss of grafts by breaking can be lessened by bracing and by pruning. A lath or board tied securely or nailed to the stock in a way that will permit it to extend beyond and serve as an anchor to which the growing graft may be tied is a desirable method of protecting grafts from high winds. Grafts that make unusual length growth may be headed back in order to reduce the strain at the point of union. It

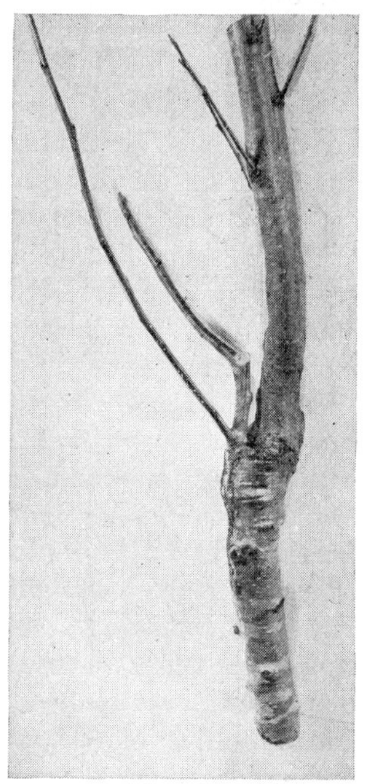

Fig. 96. Cleft graft with one scion allowed to grow and the other retarded in growth. The growth of the smaller scion at left aids in healing the stub.

is especially important that this be done on fast-growing grafts before they begin the second season of growth.

Stock in which grafts are inserted should be covered with a protective coat of a substance that will keep moisture from soaking into the exposed cut surface. Ordinary white lead paint is sometimes used, but the oil in it seeps into the wood, leaving the powdered material, which affords little protection, on the wound surface. Grafting wax is also used and is quite satisfactory. Various commercial preparations are available which, when applied properly, afford protection from moisture and tend to minimize the injurious effects of decay-producing organisms.

Squirrels frequently do considerable damage to young grafts by gnawing the tender bark and partially or completely girdling the limbs. Guards made of tin and placed around the trunk of the tree prevent squirrels from reaching the grafts, provided no other trees are nearby.

Bridge Grafting. Bridge grafting is not a means of propagation in the sense that the other methods are. It is a means of repairing tree wounds caused by cultivating implements and by mice, rats, rabbits, and other rodents. Most injury caused by animals is done during the wintertime, when other plants on which they feed are scarce. These pests are especially troublesome when snow is on the ground and their food supply is even more limited. It is not uncommon for trees to be partially or completely girdled from the ground to a height of 12 or 15 inches. Trees from nursery size to those 3 or 4 inches in diameter are subject to their attack. They do not seem to have marked preference for any one kind of tree; in one instance, rabbits injured trees of peach, plum, apple, pear, and pecan, seemingly without favor. Field rats have been known to do serious damage to bearing fig trees by girdling the trunks. Perhaps the wisest course to follow in combating such damage is to adopt preventive measures. Precautions should be taken to avoid injuries that result from cultivation, such as plow cuts and bruises. Clean cultivation discourages habitation of rodents in the orchard and thereby lessens the chances of damage. When injury is anticipated, protection may be provided with chemical repellents or by tree guards made of paper or wood veneer.

Another rather common cause of injury at or near the ground line is cold. In many cases the tissue at the base of the tree is the most tender to cold, because of its later maturity; and frequently the temperature of the air at this point, particularly on a still night, may be several degrees lower than it is even 2 to 3 feet higher.

Trees that have been injured to the extent of having all the tissue outside the xylem destroyed, even for a short distance, are said to be *girdled*. Those that are completely girdled die of root starvation sooner or later unless the connection between the root system and supporting top is reestablished. Some kinds of trees callus-over wounds, either by regeneration or overwalling, and rebuild the destroyed phloem tissues before their root systems cease to function. Elm trees are difficult to kill by girdling because of their ability to callus-over wounds rapidly. A pecan

FIG. 97. Lateral wound on apple showing bridge grafts which have united successfully and completed one season of growth. (*Courtesy of Dept. of Hort., Mich. State Coll.*)

tree is known to have callused-over about 5 feet of girdled area, but such instances are extremely rare. Fruit trees will seldom callus-over large wounds that completely encircle the trunk.

Sap flow between the upper and lower margins of a girdled area may be reestablished by causing congenial scions to unite with the tree both above and below the girdle, in such a way as to bridge the gap. This operation is called *bridge grafting*.

It should be done in early spring about the time the injured tree starts growth. Only thoroughly dormant scions are used. The irregular edges of the girdled area should be cut back evenly to fresh tissue preparatory to inserting the bridge grafts. The scions are cut at each end; after the matrix is prepared above and below the injury on the stock, as for the inlay or the bark graft, they are inserted, tied, and waxed. Large wounds may require more than one bridge graft. It is customary to space them 2 or 3 inches apart. After they are united, they transport elaborated sap past the injured area to the roots.

Approach Grafting. The approach graft is a special form of grafting in which the scion unites with the stock while it is still attached to the parent plant. It is therefore necessary that the plants to be used for stocks and as a source of scions, respectively, grow close together. The stock may be moved in a clay pot or it may be transplanted bare-rooted to the vicinity of the tree from which scions are to be obtained. On both the stock and scion a long cut is made through the cambium and slightly into the wood. The cut surfaces are brought together, and the stock and scion are tied firmly and waxed. After the two have united, the scion in some cases is severed below the union and the stock above that point, resulting in a new plant that is composed of a rootstock and a grafted top. In other cases, two parts of a plant, or parts of two plants, are caused to unite and grow together permanently, and neither component is severed below the point of union. Approach grafting is also called *inarching*.

The technique of either the bark graft or the inlay graft may also be used for approach grafting. Though the approach graft is tedious and cumbersome, it has a variety of important uses.

Difficult Species. Approach grafting is used in the multiplication of plants that are extremely difficult to propagate by other asexual means. It may be used, for example, in propagating the Muscadine grape, which responds very poorly to ordinary methods of graftage, and is also used in propagation of avocado and mango.

Novelties. Approach grafting is used to some extent, principally in European countries, as a means of providing novelties for ornamental gardens. At one place in Italy the tops of oaks on each side of an avenue have been grafted together by means of inarching.

Bracing. Living braces may be provided in the framework of a tree by grafting a small side limb from one branch into an adjacent branch. The same protection may be provided by twisting together small limbs from two branches and letting them

FIG. 98. A living brace designed to strengthen the weak crotch shown at X. The lateral limb arising at Y was caused to unite with the limb shown on the right at Z. A spiral inarch was used; it was made in April, and the picture shows it as it appeared the following winter.

unite naturally. These are valuable in strengthening limbs that form V or otherwise weak crotches. Living braces grow and become stronger as the tree becomes larger and may ultimately provide a more effective support than could be provided in any other way.

Fig. 99. The brace graft, shown in Fig. 98, after 20 years, the brace being shown at X.

Fig. 100. Living braces of apple created by twisting branches from adjacent limbs together and allowing them to grow together. (*Courtesy of Dept. of Hort., Mich. State Coll.*)

Repairing Trees. Approach grafting may also be used to repair trees that have been girdled or injured in some way near the ground line. Young trees are planted about the base, and their tops are inarched into the trunk above the injured area. These ordinarily make a very rapid growth and ultimately become a permanent part of the old tree. A sprout that arises from below

Fig. 101. The two small trees on either side of the center one were inarched into it above the girdled area, shown at *X*.

a wound on the trunk of a tree may be inarched into the trunk above the wound; when it unites, it provides a connection that will cause more even and rapid healing of the wound.

Replacing Diseased Rootstocks. Citrus trees on roots damaged by footrot or gum disease have been successfully inarched onto resistant sour orange seedlings planted around the base of the tree.

Grafting Wax. A suitable grafting wax can be made by mixing, by weight, 10 parts of pine resin, 2 parts beeswax, and 1 part talc. These are mixed by heating and are applied in a melted condition with a ½-inch brush. The melting point of this mixture is about 165°F., and the mixture should be used when it is at about this temperature or only slightly higher.

Melted paraffin is suitable as a waxing material and is used quite commonly for early spring budding and grafting. It is

Fig. 102. A good type of wax melter on right and thermos bottle for using melted paraffin on left.

not suitable for budding or grafting when the temperature is above 85 or 90°F., because of its tendency to melt and to cause scalding of the bud or graft. A pint Thermos bottle shown in Fig. 102, with a brush handle inserted through a cork stopper, is a very convenient type of dispenser for melted paraffin.

QUESTIONS

1. Define grafting. Distinguish between grafting and graftage.
2. When is the season for cleft grafting? What size stocks are suitable for the cleft graft?
3. Describe the steps in preparing the stock and the scion for the cleft graft.
4. What are the objects of inserting two scions in one stock? What procedure is followed in the event both grow?

5. Describe the steps in preparing the stock and the scion for the whip graft. What size stocks are suitable for this method?

6. What is the distinction between bench grafting and nursery grafting? When is the season for each?

7. Tell how to prepare the stock and scion for the bark graft. Distinguish between a terminal and a side graft.

8. Describe the inlay graft.

9. When is the season for bark and inlay grafting? Why is it different from the season for other methods of grafting?

10. What treatments are involved in the aftercare of grafts?

11. What are the chief uses of the veneer graft? Describe the cutting graft.

12. What are the different uses of the approach graft?

13. What various treatments are prescribed for trees that have been girdled?

SUGGESTED REFERENCES

Beaumont, J. H., and Ralph H. Moltzau: Nursery Propagation and Topworking of the Macadamia, *Hawaii Agr. Expt. Sta. Circ.* 13, 1937.

Brown, Gordon, G. A.: Method of Topworking Pear Trees for Early Maximum Production and for Reducing Stony Pit Loss, *Oregon Agr. Expt. Sta. Bull.* 438, 1946.

Bryant, L. R., and George Beach: Propagation of Plants, *Colo. Agr. Expt. Sta. Bull.* 468, 1941.

Cardinell, H. A., and F. C. Bradford: Grafting in the Apple Orchard, *Mich. Agr. Expt. Sta. Bull.* 142, 1934.

Cooper, William C., and Edward O. Olson: Influence of Rootstock on Chlorosis of Young Red Blush Grapefruit Trees, *Proc. Am. Soc. Hort. Sci.*, **57**:125–132, 1951.

Hansen, J. C., and E. R. Eggers: Propagation of Fruit Plants, *Calif. Agr. Ext. Serv. Circ.* 96, 1951.

Peck, G. W.: Topworking and Bridge Grafting, *Cornell Univ. Agr. Ext. Serv. Bull.* 154, 1936.

Roberts, R. H.: Top and Double Working Apple Trees, *Wisconsin Agr. Expt. Sta. Bull.* 432, 1936.

Snyder, John C., and Richard D. Bartram: Grafting Fruit Trees, *Wash. Agr. Ext. Serv. Bull.* 442, 1950.

Talbert, T. J.: Propagation of Fruit Trees by Budding and Grafting, *Missouri Agr. Exp. Sta. Circ.* 343, 1950.

Whitehouse, W. E.: Budding and Grafting, *Natl. Hort. Mag.*, **33**: 25–36, 1954.

CHAPTER 13

Methods of Budding

Budding is a form of graftage in which only the buds at one node of a scion are inserted in a stock. Buds are cut from scions in such a way as to have a relatively small amount of bark surrounding them; and in some methods a small sliver of the wood beneath the buds is included also. Buds are customarily placed on the stock at internodes; it is not necessary, in fact less desirable, to insert them at places where native buds grew on the stock.

Shield, or T, Bud. The most widely used method of budding is the shield, or T, bud. It is known as *shield* budding because the bud, when cut from the bud stick, resembles a shield in shape and as *T* budding because the two cuts made on the stock intersect so as to form a T.

A perpendicular cut 1 inch long, or less, is made on a smooth portion of the stock. It is followed by a horizontal cut across the top at right angles to it. These two cuts extend only through the bark and should not go into the wood.

The bud is cut by starting ½ inch below it and cutting upward and obliquely inward to a point about ½ inch above the bud. The knife is then withdrawn and a horizontal cut through the bark only is made ½ inch above the bud. Next, with a sidewise twist the bud is gently removed from the bud stick without wood adhering; this practice is known as *flipping* or *popping* the bud. The pointed lower portion of the bark is inserted underneath the two flaps of the T cut on the stock, which may have been loosened with the knife. The bud is then pushed downward until the top of it is well below the horizontal cut on the stock. When the bark of the stock slips readily, the bud can be pushed into place with little difficulty.

Buds may also be cut in about the same way as just described, except that the horizontal cut extends into the wood, which is transferred along with the bud. If the bark on the scion will not slip, it is necessary to cut the bud with some wood adhering. If the bark of the stock does not slip well, it may be desirable to cut the bud so as to include the sliver of wood, since it makes the bud more rigid and easier to insert in the stock. Under no circumstances should an individual bud be cut from

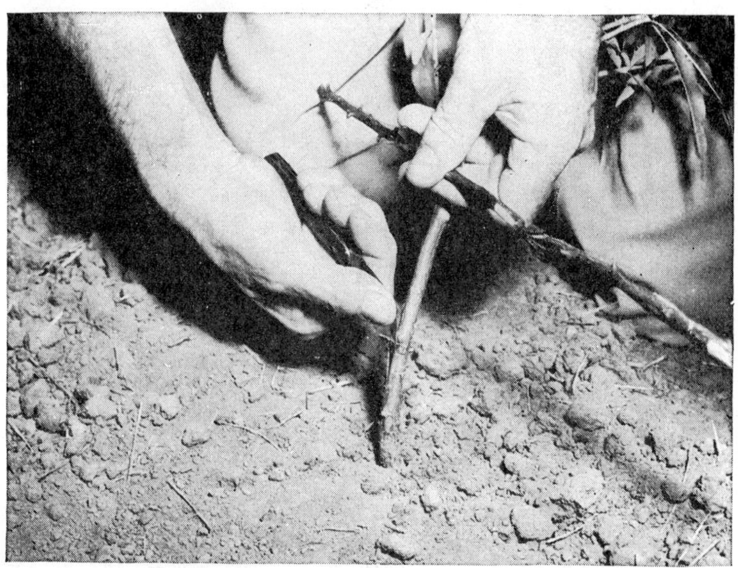

Fig. 103. Method of inserting a T bud.

the scion until the stock has been prepared for its reception so that the transfer can be made immediately.

During the seasons when, or in regions where, rainfall is excessive, *inverted*-T buds are frequently used. The horizontal cut on the stock is made at the lower end of the perpendicular cut, and the bud is cut from above rather than below. The bud may be cut with or without wood adhering, but it can be placed more easily if wood is included. It is inserted by being pushed upward beneath the flaps of the inverted T cut, which serves to shed water.

Rubber budding strips are currently the most popular for

tying T buds. Moist raffia, waxed tape, and cotton twine are also used. If the tying is done carefully, it is not necessary to apply wax. Stocks to be T-budded should not be greater than 1 or less than ¼ inch in diameter. Those the size of a lead pencil or slightly larger are considered ideal.

The T bud is the most popular of the methods of budding. It is used commonly for propagating plants, such as roses,

Fig. 104. Citrus stock cut back to bud union and budded top staked.

peaches, apples, and citrus, which are easy to bud; but it is seldom used on species that are difficult, such as the pecan or walnut.

T budding is done successfully at any season of the year when the bark of the stock will slip freely. By one practice, buds are inserted in late summer; they unite with stocks but remain dormant over the winter and are forced into growth the following spring. This practice is referred to as *dormant* budding.

After the buds grow one season, the trees that they produce are known as *one-year-old* trees, though the rootstocks are two years old. Buds may also be set in early summer and forced into growth shortly afterward. This is commonly known as *June* budding, because much of the budding is done during the

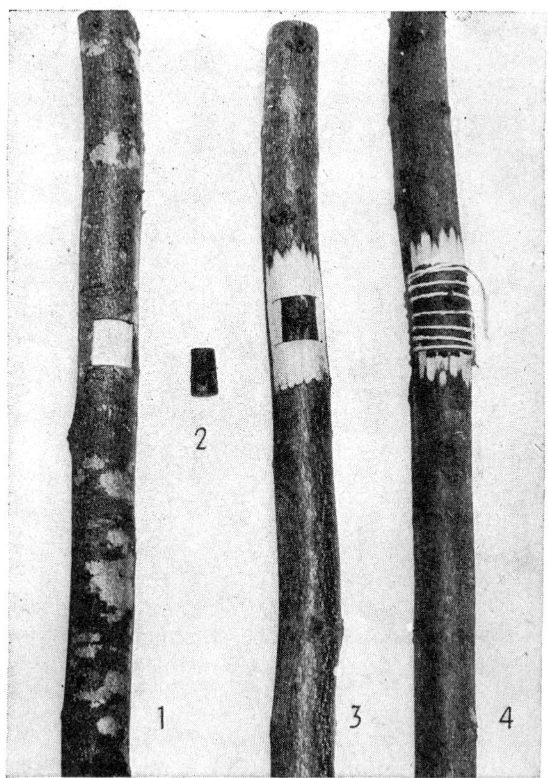

Fig. 105. Showing steps in inserting a patch bud in thick bark. (1) Stock with square removed; (2) bud; (3) bud in place; (4) bud protected with waxed patch and tied with cotton twine.

month of June. In regions where spring comes early and the growing seasons are long, June-budded plants ordinarily become large enough by the end of the first growing season to be planted out in the orchard.

Patch Bud. The patch bud is a popular form of budding in the propagation of species that are rather difficult to propagate. Essentially, the patch-bud method consists of removing a square or rectangular piece of bark from the stock and inserting in its place a bud of a desired variety on a similarly shaped piece of bark. Two parallel cuts, ¾ to ⅞ inch apart, are made on the stock, preferably with a two-bladed budding knife. The cuts should be made perpendicular to the stock and should be about 1 inch long. With a sharp pocket knife, two longitudinal cuts are next made. They likewise should be about ¾ to ⅞ inch apart, and each should intersect the two horizontal cuts, result-

Fig. 106. Cross section of pecan stem through patch bud that had united and grown part of one season. Union of the patch and stock took place along the line indicated by X.

ing in the square or rectangular "patch." Similar cuts are made above, below, and on each side of a bud on a bud stick with the same tools that were used on the stock. Care should be exercised in removing the bud from the bud stick in order to avoid splitting the bark beneath the bud. The bark should be lifted carefully on one side, or both sides if necessary, and the bud loosened by a lateral twist. The bud is held in place on the scion while the patch of the stock is flipped off, and the bud is then quickly transferred to its place.

In making the transfer it is important that the delicate cambium cells on the underside of the bud and on the exposed surface of the stock be subjected as little as possible to mechanical injury and exposure to air. The bud should fit snugly in its new location and should be tied immediately. It is then tied and made airtight with cotton twine and paraffin, waxed tape, wide rubber budding strips, and other similar materials.

Stocks that range in size from ½ to 4 inches in diameter may be patch-budded quite successfully. For the larger ones it is usually necessary to pare the rough outer portion of the bark down to the thickness of the budwood bark at the time the bud is put in place. This precaution is essential to the success of the inserted bud in that it allows the pressure of the tying material to be exerted on the bark of the bud rather than on the thick shoulder of bark on either side of it. Buds for larger stocks should be selected with special care. They should be taken from smooth, straight bud sticks, and only large, plump buds should be used; small buds are difficult to force and should be discarded.

Patch buds may be inserted successfully at any season of the year when the bark will slip freely. Those that are set early in a season are usually forced promptly, while the ones that are set late remain dormant over winter and are forced the following spring.

There are several other methods of budding in which the buds unite with the stock in much the same manner as the patch bud and differ from the patch bud only in minor details. These are conveniently considered in connection with patch budding.

Ring Bud. Ring budding differs from patch budding in that a cylinder of bark is removed from the stock in order to form a matrix; and the bud, when placed, extends nearly if not all the way around the stock. The stock is completely girdled, and, if the bud fails to unite, the top part of the stock ultimately dies. The nature of the method renders it impractical except for small stocks, those not more than ⅔ to ¾ inch in diameter.

H Bud. A modified patch bud known as the *H Bud* differs from the patch bud in that the two horizontal cuts on the stock are intersected by only one longitudinal cut. The two flaps of bark on either side of the longitudinal cut are lifted slightly, and the bud patch is inserted underneath, from above or below. In preparing the bud, the two horizontal cuts are made and the sides cut so as to form a square; but the longitudinal cuts are

Fig. 107. Showing one season's growth of patch buds inserted in rough bark of pecan. The shoots are braced to prevent them from being broken by wind. (*Courtesy of Tex. Agr. Exp. Sta.*)

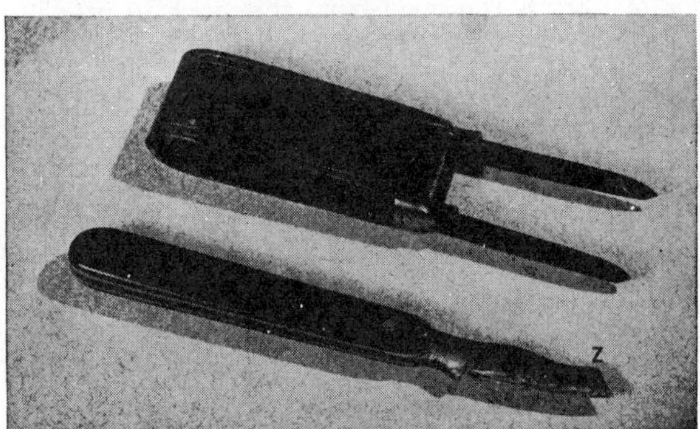

Fig. 108. A good type of budding knife above with projection at Z for use in inserting T buds. Two-blade patch-budding knife above.

cut at an angle of about 45 degrees, to enable the bark of the stock to fit over them more snugly.

Skin Bud. The skin bud is a comparatively new addition to the arts of plant craft. The matrix on the stock is prepared by cutting just through the bark, but not into the wood, in such a way as to remove an oblong piece of bark $\frac{2}{3}$ inch long. The size of stock determines, in a measure, the width of the oblong

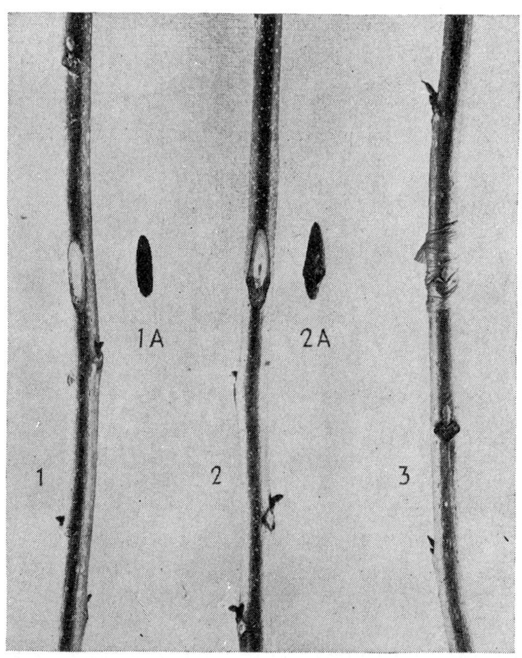

FIG. 109. Steps in skin budding. (1) Stock; (1A) sliver of bark removed from stock; (2) bud stick, and (2A) bud removed from node; (3) bud applied to stock and tied with rubber budding strip.

piece of bark removed. In a similar way the buds at a node are cut, beginning $\frac{1}{3}$ inch below the bud, cutting inward, then upward, between the bark and wood, and out again $\frac{1}{3}$ inch above the bud. The thin sliver of bark and bud or buds is applied to the matrix, tied with rubber bands or with cotton twine, and sealed with melted paraffin.

At present, the principal use of the skin bud is in the budding of small nursery pecan trees. Doubtless it could be used upon hickory nut and walnut also. Skin budding may be done from early spring until late fall. Current-season buds may be used; or previous-season buds may be used from storage or direct from the tree as needed during the budding season.

Chip Budding. Budding by this method may be done at seasons of the year when the bark does not slip. The usual season for chip budding is a period beginning about 2 weeks before active growth starts in the springtime and continuing for about 5 or 6 weeks. Nurserymen sometimes resort to chip budding in

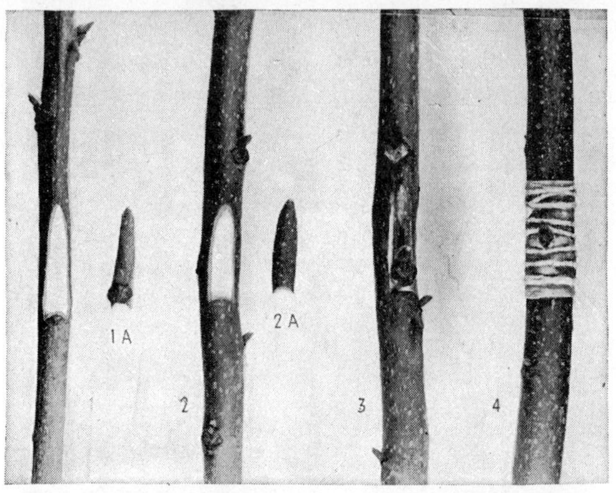

FIG. 110. Showing different steps in inserting chip bud. (1) Scion; (1A) bud; (2) stock; (2A) chip removed from stock; (3) bud in place on stock; (4) bud protected with waxed cloth and tied with string.

the summer when drought causes stocks to become partially dormant. Under such conditions the T and patch buds cannot be used effectively, but the chip bud can be used provided the stocks are not too thoroughly dormant.

It is important to select a smooth place between nodes on the stock at which to insert the bud. First a long downward cut is made through the bark and slightly into the wood of stock. The cut should be 1 to 1¼ inches long and should constitute a smooth plane. A second downward cut is made so as to intersect the lower portion of the first cut at an angle of about 45 degrees. As a result of these two cuts a chip of bark and wood is removed from the stock.

The bud from the bud stick is removed in the same manner, beginning the first cut ¾ inch above the bud and the second ⅓ to ½ inch below it. It seems to be important that the bud be on the lower and, consequently, thicker part of the chip. The chip is removed from the stock and the bud slipped into its place so as to make a close fit. Generally the bud will not fit perfectly on both sides; it is important, however, that the two cambium layers coincide on one side at least. Chip buds should be tied securely in place; if cotton string or raffia is used for tying them, wax, melted paraffin, or waxed patches should be applied to prevent drying. If waxed tape, rubber budding strips, or budding tape are used properly, no other protective material is required for the bud.

For chip budding, the stock should not exceed 1 inch in diameter and preferably should not be larger than ¾ inch. Budwood from which buds are secured should be thoroughly dormant at the time the budding is done.

Chip-budded trees have a long growing season and are ready for planting the fall following the spring during which they were budded. The system injures the stock but little and, if failure occurs, some other method may be used during the same or a following season.

QUESTIONS

1. In what respects is budding different from grafting?
2. When is the favorable season of the year for T budding? Patch budding? Chip budding?
3. What is meant by dormant budding? June budding? Flipping the bud? Inverted T bud?
4. Under what circumstances is the patch bud used instead of the T bud?
5. What are the merits of the chip-bud method?
6. What kinds of budwood are used for T budding? Patch budding? Chip budding? Skin budding?

SUGGESTED REFERENCES

Camp, A. F.: Citrus Propagation, *Florida Agr. Ext. Serv. Bull.* **139:** 9–23, 1950 rev.

Gould, H. P.: Apple Growing East of the Mississippi River, *U.S. Dept. Agr. Farmers' Bull.* **1360:**8–9, 1924.

Hansen, C. J., and E. R. Eggers: Propagation of Fruit Plants, *Calif. Agr. Ext. Serv. Circ.* 96, 1951 rev.

Havis, Leon *et al.*: Peach Growing East of the Rocky Mountains, *U.S. Dept. Agr. Farmers' Bull.* **2021**:10–11, 1951.

Lewis, I. P.: Grafting and Budding Fruit Trees, *Ohio Agr. Expt. Sta. Bull.* 510, 1932.

Rosborough, J. F., F. R. Brison, C. L. Smith, and L. D. Romberg: Propagation of Pecans by Budding and Grafting, *Texas Agr. Ext. Serv. Bull.* 166, 1949.

Schrader, A. Lee, and W. E. Whitehouse: The Budding and Grafting of Fruit Trees, *Maryland Agr. Expt. Sta. Bull.* 278, 1925.

Whitehouse, W. E.: Budding and Grafting, *Natl. Hort. Mag.*, **33**:25–36, 1954.

Yerkes, Guy E.: Propagation of Trees and Shrubs, *U.S. Dept. Agr. Farmers' Bull.* **1567**:38–45, 1932.

CHAPTER 14

Propagation of Important Horticultural Plants

The general methods of propagation by seeds, bulbs, layers, cuttings, budding, and grafting have been outlined in the preceding chapter. The use of these methods and the specific practices followed in the propagation of different horticultural plants, particularly fruits, involve a more detailed consideration.

Stone Fruits. The peach, plum, apricot, cherry, almond, and nectarine comprise an important group of plants known as *stone fruits*. Standard varieties of these fruits will not come true from seed. In producing nursery trees, the common procedure is to grow rootstocks from seeds or cuttings and then to bud named varieties upon them. Grafting is rarely used on these plants in the production of nursery trees, because gum forms readily at the points where grafts are inserted and interferes with the union of the stock and scion.

Rootstocks. The different stone fruits unite quite readily when interbudded or intergrafted, but the resulting unions vary widely in degree of compatibility.

For *peaches*, a popular commercial rootstock is a small peach, grown extensively in Tennessee and the Carolinas. Seedlings of this peach are known as *naturals.* A red-leaf strain has been introduced and is popular because in the aftercare of trees, shoots of the rootstock with their red leaves can be readily distinguished from the green-leaved shoots of the budded top. The natural rootstock produces a good root system, makes a strong and vigorous growth, is adaptable to a range of soil conditions, and is congenial with most varieties of peaches. It is used extensively, but unfortunately is subject to the attack of nematodes, a common pest in many peach-growing areas. The seed supply is not always plentiful, but this is being corrected by the planting of trees to provide a source of seed. Seed from named varieties of peaches are also used to produce rootstocks.* The Lovell, a popular peach for canning and drying in California, is one which is used extensively. One advantage is that the pits are always available in quantity as a by-product of the processing industries. Lovell is susceptible to nematodes but otherwise is regarded as being a very suitable rootstock. There

FIG. 111. Lovell peach seedlings of suitable size for T budding.

are several rootstocks which are used for peaches that are to be planted where a known hazard exists. The Chinese wild peach (*Amygdalus davidiana*) is definitely tolerant of alkali and is used satisfactorily as a rootstock when peaches are grown on alkaline soil. It is not adapted to conditions of poor drainage. The Shalil peach, introduced from India in 1913, and also a flowering peach, known as S-37, are definitely resistant to nema-

* Recently, own-rooted Elberta peach trees have been propagated from cuttings of parent trees indexed and certified to be free of virus diseases for two years. No previous work has given promise of uniform success of this method with peaches. This *may* represent the beginning of a new and desirable method of commercial propagation.

tode damage. The use of these and other kinds with known resistance is recommended when peaches are to be planted where the nematode pest occurs.

The same kinds of rootstocks are used for *nectarine* as for peach.

For *plums,* several different kinds of rootstocks are used. Myrobalan plum is a popular rootstock in the North and also on the Pacific coast. In New York, 15 varieties of plums, comprising

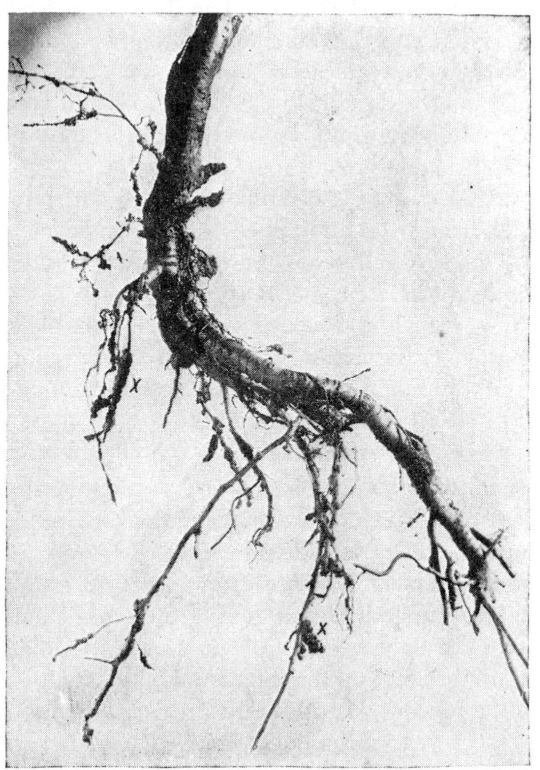

FIG. 112. Nematode injury on root of peach tree. Characteristic root-knot injury indicated at *X*.

European, Japanese, and native species, were tested for 20 years on four plum rootstocks and on peach. Results showed Myrobalan to be superior to the others tested. It will grow in fairly heavy, wet soils, and that is an advantage if plums are to be planted in such a location. Marianna plum rootstocks are used widely in the South, because of resistance to nematodes, crown gall, and oak root rot. Peach is also a popular rootstock for most varieties of plums. Trees on this rootstock are less subject to bacterial gummosis than those on the plum rootstocks. Preliminary results indicate that Shalil and other nematode-resistant peach rootstocks can be used satisfactorily where resistance is needed. Apricot is also resistant to nematode and is used successfully for plums. Almond is occasionally used as a rootstock, since it is a deep-rooted plant and is drought-resistant.

Apricot varieties are grown principally on three different rootstocks. In soils favorable to apricot roots, that rootstock is considered the most satisfactory. For wet soils and areas of high-water table, Myrobalan plum is preferred. In sandy soils, peach is used more commonly.

Three rootstocks have been widely used in the propagation of *cherries*. These are the Mahaleb (*Prunus mahaleb*), Mazzard (*P. avium*), and Morello, and there has been considerable controversy as to the relative merits of the three. It has been fairly well established that Mazzard is the most desirable stock from the standpoint of the orchardist. Trees live longer on this stock and withstand drought and diseases better than on Mahaleb. It is considered that the Mazzard rootstock is more subject to injury by winter cold than Mahaleb; this apparently is true of trees in the nursery and also of larger trees in the orchard. Seedling trees of Mazzard are more difficult to produce in the nursery, on account of leaf diseases. Mahaleb trees make better growth in the nursery and are larger and more uniform than Mazzard of the same age, but they grow more slowly and are less vigorous as they become older, and sooner or later they exert a dwarfing effect on the top. Briefly, Mahaleb makes better growth in the nursery, but Mazzard produces the better orchard trees. In recent years Morello rootstock has been used to good advantage in California in heavy, shallow, or wet soil. It produces suckers freely and these are the source of rootstocks. It has some dwarfing effect and does not make a good union with all varieties.

Handling Seed. Germination of seed of the stone fruits is influenced by the bony endocarp which surrounds the seed, and also by the rest period. The inhibiting influence of the endocarp is slight with some seeds and pronounced in others. Prior to planting, the seed should be stratified for a prolonged period to encourage splitting of the strong covering. They should also be kept at a temperature of from 34 to 40°F. while stratified to break the rest period. The length of time at the low temperature required to break the rest period varies with different stone fruits, but 8 to 10 weeks is necessary for most kinds. Since the seeds will germinate at low temperature, the rest-period-breaking treatment should be timed so that it will terminate at the usual planting season. In regions that have sufficient cold weather during the winter to break the rest period, the seeds may be stratified outdoors, or even fall-planted in the nursery row.

Growing Rootstocks from Cuttings. Rootstocks of Marianna and Myrobalan plum can be grown from seeds and from hardwood cuttings. Those from cuttings are more uniform in the various characteristics that make them useful than the ones grown from seed. Two selections that are being used quite commonly are Marianna 2624 and Myrobalan G29. Cuttings are made from dormant shoots about the size of a lead pencil; they are cut to a length of 6 to 8 inches during the winter and held in cold storage at a temperature of from 32 to 36°F. until time for planting. Shortly before the beginning of growth in the spring they are planted in nursery rows with a spacing of 8 to 10 inches. They ordinarily root readily, make good growth, and are large enough to bud the following summer or fall.

Budding. The T bud is the most common method used in propagating nursery trees of the stone fruits. There are three general seasons for such budding.

JUNE BUDDING. Where the growing season is long, peach pits that are planted in the fall, or those that are stratified over winter and planted in later winter, will germinate in early spring. With favorable growing conditions, the young seedlings that grow from them will be large enough to be T-budded in June. The young trees will be 12 to 15 inches high and ¼ inch in diameter near the ground. The buds are usually forced at the time of budding, and the resulting trees, known as *June buds,* are ready for planting the following winter, about a year from the time the seed were planted. June-budded trees often grow from 2 to 3 feet between the time buds are set in the summer and the end of the growing season in the fall; under very favorable growing conditions they may make 4 or 5 feet of growth.

DORMANT BUDDING. In regions where the growing seasons are relatively short the seedling trees are seldom large enough to be budded before late summer. They are budded mainly in August; the buds unite but remain dormant over winter and are forced into growth the following spring. Since the buds start growing in the early spring, they normally make large well-developed trees by fall. They have a two-year rootstock and a one-year top, and are sold as one-year-old, or yearling, trees.

Fig. 113. Lovell peach seedlings shown in Fig. 111, T-budded in June slightly above ground line, and tops lopped to force buds at X.

SPRING BUDDING. Rootstocks that have grown one year may also be budded in the spring of the second growing season. Such buds are usually forced immediately and they normally make sufficient growth to produce salable trees, comparable to dormant budded trees, during the remainder of the growing season.

Though the T bud is the method most commonly used in propagating stone fruits, other methods are sometimes used. If the bark on the rootstock does not slip satisfactorily, as, for example, in the very early spring, the chip-bud method may be used with good results. The whip-graft method is also used occasionally, on the plum particularly. Rootstocks that have completed one season of growth are whip-grafted at the crown. This is done in late winter or early spring. Such trees are more

Fig. 114. Seedling peach trees, Z, started from seed in early spring, T-budded in August at X, forced following spring, and formed these well-developed tops in 3 months.

likely to develop crown gall, because of the mechanical injury below the surface of the soil.

Budwood. Current-season budwood is used for June- and for dormant-budding of the stone fruits. Bud sticks about $\frac{1}{4}$ inch in diameter, or slightly smaller, are of suitable size, and wood that has attained a degree of maturity, as shown by a golden-brown color of bark, is considered good. The budwood may be

Fig. 115. Marianna plum cutting, planted in March, developed roots at base and sufficient top growth to be T-budded in June or August.

ripened by clipping the leaves several days in advance, so that the petioles will fall off before the buds are used. The common practice, however, is to cut the leaves off at the time the bud sticks are cut; a portion of the petiole is left and is useful in handling the bud and in pushing it into place in the stock, particularly with flipped buds.

Dormant scions of the previous season are used as a source of buds for chip budding and grafts for whip grafting. Most stone fruits produce some blind buds, and these, of course, are useless for propagation and should be discarded whenever they are encountered. Scions from bearing trees usually contain a large percentage of triple buds—one vegetative bud and two fruit buds—at each node. Though these are suitable for budding, they are somewhat cumbersome to handle and insert. The best scions are obtained from fairly young trees that are making normally vigorous growth, but which have not reached an age at which they form fruit buds. Trees may be cut back heavily during the dormant season to cause them to produce good budwood for use the following budding season.

Top-working. It is frequently desirable to change the variety of large trees by top-working. This may be done in several different ways. Probably the most satisfactory system is one by which the top is dehorned and the new shoots that develop afterward are budded to the desired variety. Several steps are involved.

DEHORNING. This should be done when the trees are dormant, preferably in late winter. The top should be cut back to points where the framework branches are from 1 to 2 inches in diameter. As far as possible, limbs should be cut so as to result in a dehorned tree that is symmetrical in shape.

THINNING THE YOUNG SPROUTS. A large number of young sprouts will develop on each stub of the dehorned tree during the spring. These should be thinned out when the largest are 6 or 8 inches long. Four or five of the largest and most vigorous shoots are left. It is important that one shoot be allowed to grow near the end of each stub and that all those that are left be properly spaced along the stub; on stubs that extend out in a horizontal direction, only shoots that grow out from the upper side should be selected. In thinning the competing shoots, it is not always necessary to cut them off. They may be twisted and partially broken in such a way that they will continue to support the tree and prevent sunscald; and at the same time they will not interfere with the growth of the selected shoots.

BUDDING. The shoots that begin to develop on stubs in early spring and are selected somewhat later may be budded in early summer and the buds forced into growth the same season; or the budding may be done late in the summer and the buds, which remain dormant over winter, forced into growth the following spring. Though all the selected shoots on a stub are usually budded, two successfully budded shoots on each stub are normally sufficient.

AFTERCARE. It is necessary to recut the stub back to the nearest growing shoot if none happens to grow on the end of the stub.

FIG. 116. Well-developed Bruce plum tree on peach rootstock. Arrow indicates line of union.

Native shoots may continue to develop on the old framework for several seasons. These should be cut out as often as necessary. When the budded top has grown sufficiently to provide shade for the framework, native sprouts do not develop so freely.

The cleft graft, bark graft, and inlay graft are methods that are occasionally used in top-working large trees of the stone fruits. These methods may be used with a fair degree of success, provided the grafts are set in limbs that are 1 inch or less in diameter. If limbs much larger than this are grafted, the formation of gum by the stock is likely to interfere with the union; this tends to restrict the growth of the graft and delays the healing of the wound by overwalling.

Pome Fruits. The principal pome fruits grown in the United States are the apple, pear, and quince. Of these the apple is the most popular and most widely grown. Pears are grown commercially in several important producing regions, and widely

Fig. 117. Bacterial gummosis, X, above, but not below, union of Bruce plum on Marianna plum roostock. Arrow indicates line of union. Compare with Fig. 116.

as home orchard trees. The quince is a very minor fruit crop, but it is used as a rootstock for dwarf pears and apples and ornamental quince.

Propagation of Apple. Propagation practices influence in several ways the performance of apple trees, and for this reason methods of propagation are of concern to the applegrower as well as the nurseryman.

Seedling Rootstocks. Many different rootstocks are used in the propagation of apple trees. The standard one for many years was the seedling French crab. In some cases the seedlings were grown in France and imported; in others the seeds were imported and the seedlings grown in the United States. In recent years the trend has been to plant the seeds of named varieties,

Fig. 118. One-year-old apple-seedling rootstocks and scions for whip grafting.

notably Winesap, McIntosh, Northern Spy, and Delicious, for the production of seedling rootstocks. Seeds are obtained from the apple pomace in the process of making cider. It is important that the seeds be separated from the pomace while it is fresh and before it begins to heat as a result of fermentation. The seeds are washed, stratified, and kept continuously moist to preserve their viability. In addition to this they should be subjected to a temperature of from 34 to 40°F. for 8 to 10 weeks to break the rest period. The seeds may be stored in loosely filled bags between cakes of ice, in which case stratification is omitted.

The common practice is to plant apple seeds directly into rows in the nursery, as outlined in a previous chapter. The seedlings are allowed to grow 1 year in the nursery. They are then dug and used in several different ways in the final production of nursery trees.

The quality of seedling rootstocks is determined largely by the soil in which they are grown and the treatment given them during the first year of growth. A deep, fertile, well-aerated soil encourages the growth of seedlings with long, straight, thick taproots that have few side roots. These are suitable for piece-root grafts, one root often providing 2 or 3 piece roots. Some propagators prefer whole roots with well-developed branch roots for rootstocks to be used in either budding or grafting. The formation and development of branch roots can be encouraged by severing the taproot of the young seedling 3 or 4 inches below ground. This may be done with a long, sharp knife, and irrigation should follow immediately to settle the soil around the roots. One-year-old seedling plants that are lined out in the spring for budding the following summer produce branched roots as a result of having been transplanted. Most of the apple-seedling rootstocks in the United States are grown in the Kaw River valley in Kansas and on the Pacific coast, particularly in Washington and Oregon.

Intermediate Stocks. Double-working is frequently used in propagating apples to produce straight instead of crooked or scraggly-growing trees, to prevent damage from certain diseases and insect pests, to introduce strength and hardiness into the framework of the tree, and also to produce dwarf trees. Scions of the Delicious variety are frequently used as splices at the ground line to prevent collar rot on a susceptible variety such as Grimes Golden. The Hibernal and Virginia Crab are vigorous and hardy stocks used commonly to provide cold hardiness for the trunk and the framework of trees. Astrachan is used as an intermediate stock because it is resistant to woolly aphis and is also hardy to cold. Clark Dwarf is a variety that is used as a splice to cause dwarfing of the top which may be budded upon it.

Clonal Rootstocks. Certain varieties of apples are propagated vegetatively to produce clonal rootstocks for apples. Thus Paradise, Doucin, and several strains of Malling apple are regularly propagated by mound layers. After the layers have become rooted, they are transplanted to nursery rows and are ultimately budded or grafted to produce dwarf trees. Quince, used also occasionally as a rootstock to produce dwarf apples, is grown readily from hardwood cuttings. The Northern Spy and Astrachan apple varieties are resistant to woolly aphis, which attacks the roots and crown. They do not grow well from cuttings but can be produced by grafting long scions onto short nurse roots. These grafted plants are set deeply so that roots will ultimately develop on the lower part of the scion.

Budding. Two, and sometimes three, years are required to produce an apple tree with a one-year top of standard variety. Seedling rootstocks that have grown 1 year from seed are dug in the fall or winter, lined out in the nursery row, and budded during the second season. If the budding is done in early summer, the bud may be forced to make sufficient growth the same season to be suitable for orchard planting the following winter. This practice is followed where the growing seasons are long. More often the one-year seedlings are lined out at the beginning of the second growing season and budded during the late summer. The buds remain dormant over winter and are forced the following spring by cutting the stock off 2 or 3 inches above the bud. As soon as the bud makes from 4 to 6 inches of growth, the stock is recut immediately above the growing shoot. Such trees make 4 or 5 feet of growth by fall and are known as *yearling* trees.

The T bud is the prevailing method of budding the apple. It has long been the practice to insert T buds in apple stocks slightly below the ground line. To do this it is necessary to clear the soil away and wipe the adhering soil from the stock. It has been shown that buds set 3 to 4 inches above the crown begin growth the following spring more evenly and make taller and larger trees than those inserted at a lower level. Current-season buds are ordinarily used for budding apples, though one-year-old scions that have been kept dormant in cold storage may be used as a source of buds for early budding.

The chip-bud method may also be used successfully in budding apples. It is used in the early spring on one- or two-year-old rootstocks, and the buds if forced immediately will make good growth during the following season.

Grafting. Seedling apple trees that have grown 1 year may be grafted in the nursery or bench-grafted. In either case the

Fig. 119. Two-year budded trees in nursery row. (*Courtesy of A. F. Lake, The Shenandoah Nurseries, Shenandoah, Iowa.*)

whip graft is the method used, and scions are inserted on the root below the crown.

If apple trees are grafted in the nursery, the work should be done in late winter or very early spring when temperatures are likely to be favorable. Soil is cleared away and grafts are placed on the root. Cotton string or nursery tape is used for tying; the moist soil is pressed about the scion leaving only the top bud exposed. It is not necessary to remove the cotton string since it soon decays; tape must be removed within 1 month or 6 weeks in order to prevent girdling.

Bench grafting is used more commonly than nursery grafting. The seedlings are dug in the fall and grafted indoors during the winter months. The grafts are made and wrapped with string, nursery tape, or masking tape and are then packed in moist, well-aerated material and allowed to callus for a period of 10 days to 2 weeks. A temperature of from 75 to 80°F. is favorable for this. Callused grafts may be planted out in the nursery row immediately; or if planting is delayed by unfavorable weather conditions, they may be held for several weeks in cold storage, in moist packing, at a temperature of about 32°F. In planting apple grafts in the nursery, they are spaced 8 or 10 inches apart in rows that are 5 or 6 feet apart. It is important that individual grafts be placed so that the union is slightly below the ground level. The most desirable shape of tree is produced if only one shoot is allowed to grow from the scion; others should be pruned off at necessary intervals during the first season. Such grafts may make from 4 to 6 feet of growth during the following season and are large enough for orchard planting the following fall or winter.

Fig. 120. An apple tree as it appeared one year after the scion was grafted on the seedling rootstock at X. The top of the scion at the time it was grafted on the stock is shown at Y.

The union of the stock and scion often results in callus overgrowths that are physical abnormalities. These can be prevented to some extent by wrapping the grafts with waxed tape or nursery tape instead of string. The tape will girdle the stock if it remains too long, but in the meantime it will usually have caused the stock and scion to make a smooth union. In other cases the succulent callus tissue presents a favorable medium for the entrance of organisms that cause crown gall and similar types of abnormal growth. Disinfecting stocks and scions before preparing them, and also after they are joined together, by dipping in Bordeaux mixture or bichloride of mercury solution is effective in reducing these diseases. It is well also for the propagator to sterilize his knife and hands at intervals. Despite these precautions abnormal overgrowths on whip-grafted trees are of frequent occurrence in some seasons.

Own-rooted or *scion-rooted* trees are produced by grafting a long scion onto a relatively short nurse rootstock. After the two have callused, the graft is planted so that the union is considerably below the ground, and soil is pressed about the scion. Roots sometimes develop on the lower portion of the scion during the first season. *Scion shoots* of some varieties that are caused to grow from below ground by deep planting are more likely to form roots than the old scion. Inverted nurse-roots are also used in encouraging the rooting of apple scions. The chief value of own-rooted apple trees is for use in experimental work.

Several different methods of budding and grafting are used successfully in *double-working* apple trees. One fairly common system is to whip-graft the intermediate stock onto a seedling rootstock. The intermediate stock is then usually T-budded with the variety that is ultimately to produce the top. If the purpose of double-working is to produce a hardy trunk and framework, the final budding is delayed 2 or 3 years or even longer, until the intermediate stock has developed properly. On the contrary, if double-working is used to produce a dwarf tree, the second budding operation may be performed shortly after the first one, since even a very short dwarfing splice is effective.

Top-working. Large apple trees may be readily top-worked by either budding or grafting. The more common practice is to dehorn the old tree by cutting limbs off at points where they are from 1 to 3 inches in diameter and to insert one or two cleft grafts in each stub. The apple is relatively easy to graft, and cleft grafts unite readily if the work is done with reasonable care. The bark or inlay graft can also be used successfully on limbs that have been cut back as for cleft grafting. By either of these methods the new top of a standard variety makes a vigorous growth and may replace the old top in one or two seasons.

By another practice the tree is dehorned in the winter, the young sprouts that develop on the stubs are thinned in early spring, and the selected ones are budded when they attain sufficient size in the summer. If the shoots become large enough to be budded in early summer the buds are forced out immediately. Buds that are set late remain dormant over winter and are forced the following spring.

Dwarf Apple Trees. Dwarf apple trees are being planted widely as novelties and for the commercial production of apples. They begin to bear at an early age, when the trees are small. The trees can be planted close together to compensate for the small yield of individual trees. The apples produced by them are comparable in size and quality to those from standard trees. The small size of the trees makes it easier to spray and prune the trees and to thin and harvest the fruit. Dwarf apple trees are propagated by budding or grafting standard varieties onto certain dwarfing rootstocks, or by using a dwarfing splice in a double-worked tree. Dwarfing rootstocks, mentioned previously in this chapter, are Paradise, Doucin, and Malling strains of apples, and Clark Dwarf is popular dwarfing intermediate stock.*

Propagation of Pears. The methods used in propagating the pear are almost identical with those used on the apple. It apparently forms callus freely and responds readily when it is either budded or grafted. The standard procedure in producing a tree to be planted in the orchard is to grow a seedling for rootstock and bud or graft a top of standard variety upon it.

Seedling Rootstocks. Most pear rootstocks are grown from seed. Pear seed is usually somewhat more expensive than apple seed because of the greater difficulty involved in extracting them from the carpels of the fruit. The seeds require special treatment in order to obtain good germination. It is important to keep them under moist conditions continuously from the time they are collected until they are planted; in addition they should be stored at a temperature of from 33 to 38°F., which serves to break the rest period and to prevent premature germination after the rest period is broken. The combination of moist conditions and a low temperature may be conveniently provided by

---

* A leading nursery uses a procedure involving three graftings; the tree consists of the original seedling rootstock, Virginia crab body stock for hardiness, Clark Dwarf splice for dwarfing, and the top of the variety for fruit production. Dwarfing has also been accomplished by removal of a ring of bark about 1 inch wide on the trunk, and replacing it in inverted position.

placing the seeds in loosely filled bags and storing them between cakes of ice. Seeds are commonly planted directly into the nursery rows, and resulting seedlings are ready to be budded or grafted after one complete growing season. They may or may not be transplanted to a new location at the beginning of the second growing season.

FIG. 121. Japanese pear body stock top-worked with Kieffer pear at points indicated by an X.

The pears used for rootstocks can be classified in three groups. *French pear* seedlings (*Pyrus communis*) from imported seed or domestic varieties, such as the Bartlett, have been the most common source of rootstocks for pears. They are adapted to a diversity of soils and are also compatible with most varieties of commercial pears. Unfortunately, this rootstock is extemely susceptible to a serious disease known as pear blight, which attacks and often destroys the roots, stems, leaves, flowers, and fruit. French pear seedlings produce suckers freely and through these, roots become infected. Several *Oriental* species are being used as pear rootstocks. They are generally more resistant to pear blight and woolly aphis but not entirely satisfactory in other respects. Trees of certain species, notably, *P. serotina* and *P. ussuriensis,* have undesirable habits of growth; most of them have heavy thorns and undesirable frameworks for top-working. Certain French varieties when top-worked on these stocks produce fruit affected with a physiological disease known as *black end.* The Callery pear (*P. calleryana*) is a popular rootstock for Kieffer and other hybrid pears in the South, because good unions are formed, and the trees make vigorous and uniform growth. Seedling trees of this species are resistant to blight and woolly aphis, and they endure drought remarkably well. It is satisfactory for French pears in Oregon, but produces weak-growing trees, and is not satisfactory as a rootstock in New York and Michigan. *Hybrid* pears, such as Kieffer, produce good seedling rootstocks, but they are not so resistant to fire blight as the Oriental species and are no better than the French pears in other respects.

Budding and Grafting. Seedling trees that have grown one season may be (1) dug and bench-grafted, using the whip-graft method, during the first dormant season; (2) whip-grafted in the nursery in late winter preceding the second growing season; or (3) budded during the second growing season either in their original location or after having been dug and lined out in a new place.

One-year-old pear seedlings that are whip-grafted, either indoors or in the nursery, usually make strong and stocky trees during the following growing season. Those that are grafted indoors are usually stored for 10 days to 2 weeks in moist insulating material at from 75 to 80°F. to encourage the formation of callus before they are lined out in the field. If budding is practiced, the seedling trees are budded during their second year of growth. Where growing seasons are long, T buds, and rarely chip buds, may be set in early spring, and the budded tops, if properly forced, may make as much as 4 or 5 feet of growth during the following season. Where growing seasons are relatively short, buds are usually inserted in the summer of the second year but not forced until the following spring. With proper aftercare the following year, the budded tops make strong and stocky trees suitable for orchard planting.

Crown gall, mushroom root rot, fire blight, woolly aphis, and nematodes are hazards to be considered in propagating the pear. Close attention to the choice of rootstocks and methods of handling will aid in reducing losses from these agencies.

Top-working. Pear trees may be easily top-worked by (1) cleft grafting, (2) bark or inlay grafting, and (3) dehorning during the dormant season and budding on young sprouts the following summer. Details of these operations and aftercare are practically the same as outlined for the apple.

Top-working is practiced much more extensively on pear trees than on most other fruits in (1) changing trees of poor varieties to more desirable kinds and (2) producing trees with a strong and hardy framework or one that is resistant to pear blight.

Double-working. Double-working is used on pears to provide for resistance to blight, for hardiness, and for compatible stock and scion combinations. In research investigations with pears through the years, a few seedling French pears have been found that are resistant to blight. Farmingdale, P18, and Old Home are examples of varieties that have such resistance. In Oregon, crosses between P18 and Farmingdale were found to be completely resistant to blight infection of the roots. The varieties P18 and P87 were found to be resistant trunk and framework stocks. Thus in the propagation of a blight-resistant tree, a common procedure is to grow seedling rootstocks of P18 $\times$ Farmingdale and to bud these with either P18 or P87 to produce the trunk and framework. After a well-branched framework has been developed, it is top-worked with the desired commercial variety.

Pears are frequently budded onto quince. Some varieties unite very poorly with it, however, and double-working is a convenient means of producing a good tree by using an intermediate stock that is compatible with the quince rootstock and the selected variety. The Hardy and Old Home varieties form satisfactory unions with quince, and varieties that do not unite well may be double-worked on a splice of one of these.

Quince Rootstocks for Pears. The quince is used commonly as a rootstock for named varieties of pears to produce *dwarf trees.* These rootstocks are grown entirely from hardwood cuttings of the Angers type of quince. Most varieties of pears form a ready initial union with quince. With all, however, breaks occur each season in the cambium region where the stock and scion meet. Thus, the two do not form a perfect union and the resulting pear tree is dwarfed. Some varieties form such poor union that double-working, as outlined above, becomes necessary to effect a satisfactory union. Dwarf pear trees are more resistant to blight by virtue of making a less succulent growth than standard trees.

Quince rootstock is tolerant of wet soil conditions and is often used as rootstock for pears in low wet spots in pear orchards. It is also used as a nurse root to encourage scion rooting of a blight-resistant variety, such as Old Home.

Hardwood Cutting. Several varieties of hybrid pears, such as Garber, Kieffer, and particularly Pineapple, can readily be propagated from hardwood cuttings. The cuttings are made from mature, one-year-old wood, during the winter. They are planted directly into field rows in late winter or early spring. Deep sandy loam soil, well supplied with moisture, is favorable for the rooting of these. They will develop into trees that are 4 feet or more in height and will be ready to transplant at the end of one season.

Fig. 122. Quince cutting, Z, produced roots at X, and top growth from a lateral bud at Y. After 1 year it was T-budded with pear, and the bud made approximately 36 inches of growth within 2 months as shown in plant at A.

Grape. The grape can be propagated in several different ways —from seed, from cuttings, by layering, and by budding and grafting.

Seed. The principal reason for growing grape plants from seed is to develop new varieties. From a group of seedlings there is the ever-present possibility that one of superior and outstanding qualities will be found. Such popular varieties as the Concord, Carman, and many others have originated as seedlings. Rootstocks can also be grown from seed but most of these are grown from hardwood cuttings. Because of the extreme variability of seedling plants of grape, commercial viticulturists never grow a vineyard from seed.

Grape seeds are small and require special treatment for best results. They may be planted in the fall and allowed to remain in the soil over winter. More commonly they are stratified in sand or peat over winter, kept at a temperature of 32 to 40°F. to break the rest period, and planted the following spring. In either case they germinate when climatic conditions become favorable. Best results will be obtained if they are planted in a well-prepared seedbed where the young plants can be given special attention with regard to cultivation, watering, and pest control during the first year. After this, they are planted in the vineyard with ample space for normal growth.

Cuttings. Most species of grapes can be grown from cuttings. Some varieties root more readily than others, and certain ones produce better root systems than others. Cuttings for propagation may be taken at any time the plants are dormant. They are ordinarily made from 10 to 15 inches long, the lower end being cut below a bud and the upper end 1 inch above a bud. Canes that are one season old, medium-sized, well-matured, and have short internodes are preferred. If the cuttings are made during the winter, they may be held in cold storage until planting time in the spring. Cuttings respond readily if planted in a well-drained sandy loam soil. They are planted in rows 4 or 5 feet apart and given a spacing of from 6 to 10 inches in the row. Individual cuttings are planted deep, leaving only one bud above the surface; to do this it is often necessary to set the cutting in a slanting position. Grape cuttings are used (1) as a means of propagating new plants of certain varieties for planting in the vineyard without being budded or grafted and (2) to provide rootstocks that may be budded or grafted to standard varieties. Normally cuttings will make sufficient growth in one season to be used as vines for planting in the vineyard; they may, however, be allowed to grow two seasons in the nursery before they are transplanted. Cuttings that are to produce rootstocks for budded or grafted plants should be disbudded before they are planted. This is done by cutting off all except the bud nearest the top of each cutting. The treatment discourages the formation of shoots from below the graft

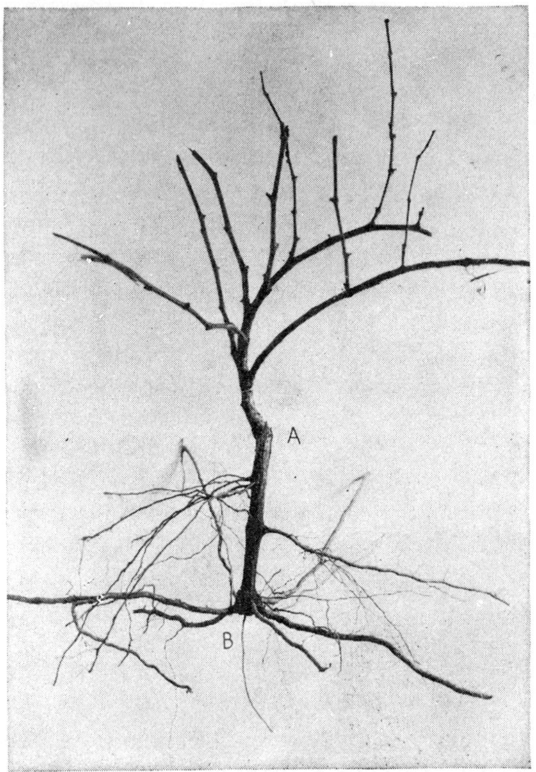

Fig. 123. One-year-old grape plant grown from a cutting. The section between *A* and *B* represents the original cutting.

union. Plants that are grown from cuttings for rootstocks are usually budded during late summer of their first season of growth or grafted during the following winter.

Layers. Certain species of grapes do not root well from cuttings. This is true of the Muscadine grape (*Vitis rotundifolia*). Since they are also difficult to graft, layers are used to propagate named varieties. Simple, compound, and continuous layers are used under varying conditions. The layers are usually made in late winter, and the rooted plants transplanted the following winter.

Grafting. There are several considerations that may prompt the grower to use grafted plants, instead of plants grown from cuttings, in establishing the vineyard. It is known that the cuttings of some varieties form poor root systems, and as a result the vines are not so productive as when grafted on some other rootstock. This seems to be true of the Delaware and Campbell, at least under some conditions. The root-knot nematode is one of the destructive insect pests of grapes. Workers of the United States Department of Agriculture found that 154 varieties of European wine grapes (*Vitis vinifera*) were very susceptible to root-knot infestation. The majority of the American species and hybrids tested were likewise found to be seriously affected when planted in nematode-infested soil. Some varieties tested have some inherent resistance to nematode injury, and this suggests their possible value for stocks. The grape rootlouse, or phylloxera (*Phylloxera vitifoliae*), is also a troublesome insect pest on all wine-grape varieties and on some American varieties. Varieties of some American species show marked immunity to phylloxera damage and are hence valuable as grafting stock for susceptible varieties. Root rot is a disease that attacks only the root system of grapes. Some species are resistant to it, others highly susceptible. These considerations and others suggest the possible value of graftage in the successful culture of grapes in the different environments in which they are grown.

The whip graft is used almost exclusively in grafting small grape vines. One-year-old stocks grown from cuttings or seed may be whip-grafted in the nursery in late winter or very early spring preceding the beginning of the second year. These will produce strong, stocky plants during one growing season and will ordinarily be large enough to be planted in the vineyard the following winter or spring. Instead of grafting the stocks in the nursery, they may be bench-grafted during the winter and replanted in the nursery in the early spring. If this procedure is followed, it is advisable to store the grafts until they form callus before planting them outside. Plants handled in this manner will likewise produce good vines in one season after having been grafted.

Quite commonly unrooted cuttings about 10 inches long are used for rootstocks. They are cut so as to have a node at the bottom and sufficient internode at the top to permit the slanting cut necessary to make the whip graft. The scion is made by cutting ½ inch above a bud and leaving sufficient smooth wood

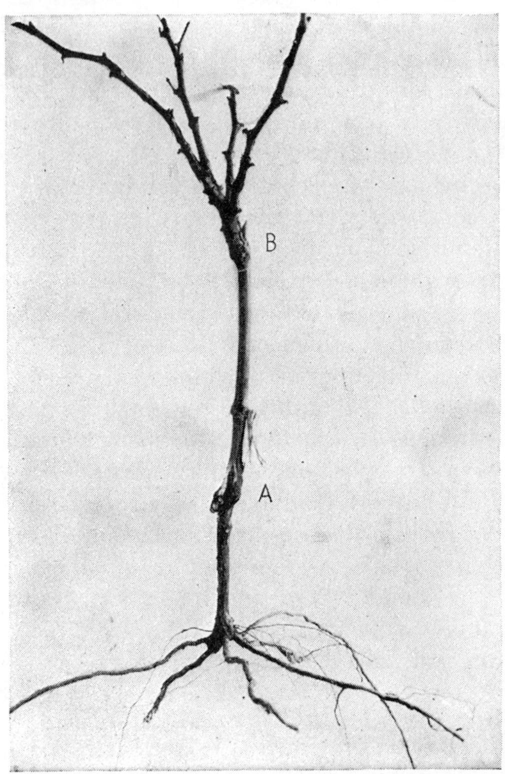

Fig. 124. Grape plant after one season of growth, produced by grafting a scion on an unrooted cutting. The point of union is shown at *A*, and the terminal end of the scion at the time it was grafted on the stock is shown at *B*.

below to make the slanting cut. These cutting grafts are usually made indoors near the end of the winter season. The finished grafts are commonly tied into bundles and packed in moist sphagnum moss, sawdust, or other insulating material and held at a temperature of from 75 to 80°F. for 10 days or 2 weeks. During this time they form the callus that is essential to a union.

Grafts that have callused may be held in cold storage until a convenient time for planting in the nursery. The planting, however, should not be delayed past the usual season when grapes begin growth. Some propagators follow a practice of planting the grafts immediately after they are made, even though it may be several weeks before the growing season. This practice is open to the criticism that the temperature in the nursery may not be favorable for callusing.

In lining out the grafts they are placed in a deep furrow with the unions at the level of the ground. A straight-edged board is used to get them all at the same level. They are placed 8 inches apart in the furrow and the soil pulled in around them. As soon as possible, the earth is mounded along the row so as to cover lightly the top of the graft. It is important that the soil be kept continuously moist around the graft union and scion, by irrigation if necessary. When the new shoots have grown through the soil and the cutting stocks are apparently well established, the soil is removed to the level of the union, and all roots originating from the scion are cut off. Any shoots arising from the stock are likewise removed.

Old established vines are frequently top-worked by use of the cleft-graft method. They are cut off at a smooth place near the surface of the ground. The grafts are inserted and moist soil is pressed about them in such a way as to leave only the top bud of the scion exposed. For stocks over 1 inch in diameter, two grafts are used, and both are allowed to grow if they unite. This may be done in late winter before the period when sap flow is excessive or for 2 or 3 weeks after that period, but grafts seldom grow if the grafting is done during the period when vines will "bleed."

Budding. Small grape rootstocks may be top-worked by budding instead of grafting. One-year-old plants grown from cuttings are usually used for budding stock. These are dug up and prepared for planting in a temporary place or in the vineyard. The preparation consists of cutting off some side roots, leaving only those that arise near the base, and removing all buds from the original cutting to discourage the growth of suckers. Lateral shoots near the top of the cutting should be pruned back to two buds each. This rootstock is then planted so that 2 or 3 inches of the old cutting is above the soil level. Soil, however, should be rounded about the base of the cutting in order to protect it from the sun until it is budded.

The chip bud is the method commonly used when grapes are budded. Late summer is the proper time for this practice. The buds are taken from budwood of the current season's growth, which has reached a stage of maturity that is indicated by the brown color of the wood and buds. The buds are inserted slightly below the ground level on the part of rootstock represented by the original cutting. They are wrapped with rubber budding strips, covered with soil to a depth of about 6 inches, and left without further treatment until the following season. In the meantime the bud is expected to form a union with the stock. When the vines start growth the following spring, the mound of soil is removed, and the stock is cut off 1 to 2 inches above the chip bud; and roots that may have formed from the bud are trimmed off and tying strips are cut also. After this the bud is again covered 1 or 2 inches with loose soil. When the shoots from the buds have made 8 or 10 inches of growth, the soil is removed finally from about the bud union and all side roots are pruned off. Plants propagated in this manner ordinarily come into normal bearing by the fourth year.

Rootstocks. The grape phylloxera, nematode, and root rot are pests that must be considered in selecting a stock for grapes. In addition, rootstocks respond differently in different kinds of soils. Rootstocks with pronounced resistance to phylloxera have been selected from *Vitis rupestris*, a common one being the so-called "Rupestris St. George." This stock has soil limitations, however, thriving in a deep permeable soil, but not in shallow, poorly drained, or clay soils. Many other stocks are known to be resistant to phylloxera, but their use is likewise complicated by soil limitations, difficulty in rooting cuttings, and difficulty in grafting. Nematodes attack most varieties of Vinifera grapes and the majority of American species. Varieties of certain species appear to have some inherent resistance to nematode injury. Most promising stocks tested are the Dog Ridge, Barnes, De Grassett (*V. champini*), and Doan (*V. doaniana*). Resistance to phylloxera is no indication of resistance to nematodes; the phylloxera-resistant Rupestris St. George is highly susceptible to nematode injury. The Dog Ridge, Champanel, and La Pryor are three stocks recommended as being resistant to root rot.

Tree-nut Crops. The walnut, pecan, and related species are propagated principally by budding and grafting named varieties onto seedling rootstocks. They are more difficult to bud and graft successfully than the stone and pome fruits. This is probably due partly to the inhibiting influence of tannin which they contain; and also to the relatively slow formation of callus, with the result that the scion is more likely to become weakened from respiration and drying before union is accomplished. It has been shown experimentally that they can be grown from cuttings and also from layers.

Propagating Walnuts. The English or Persian, black and Japanese are different kinds of walnuts grown in the United States. Of these, the English walnut is by far the most important.

Rootstocks. Several different rootstocks are used for English walnuts. The Northern California black (*Juglans hindsii*) walnut has long been the most popular rootstock, because of its vigorous growth. Under some conditions, it is quite subject to crown rot, which causes girdling and results in death of the trees. Paradox hybrids, which are crosses between the Northern California black and English walnuts, are promising as rootstocks for English walnuts, because of their superior vigor, resistance to crown rot and root-lesion nematode, and their adaptability to heavy soil. Seedlings of English walnut varieties such as Franquette are suitable for rootstocks, but unfortunately they are quite subject to damage by mushroom root rot.

Nurserymen use eastern black walnut (*J. nigra*) rootstocks for standard black walnut varieties, such as Thomas, Stabler, and Ohio.

The promising Paradox Hybrid rootstocks are grown from continuous layers. All others are grown from seed, which are usually stratified over winter and planted in early spring. The stratification helps weaken the bony pericarp, which enables the radicle and plumule to emerge. Seeds are planted at about 8-inch intervals in nursery rows and covered about 4 inches deep.

Budding and Grafting. The patch bud is the method used commonly in propagating nursery walnut trees. The rootstocks are usually budded 4 or 5 inches above the ground line, during the late summer of the first or second growing season. The buds remain dormant over winter and are forced early the following spring. The trees are allowed to grow one or two seasons and attain a height of 5 to 7 feet before they are transplanted to permanent locations. Some difficulty is encountered in budding black walnuts, because of the tendency of the primary bud to abscise. The secondary bud often has changed into a catkin, and the third bud of the group at a node is often difficult to force on account of its small size. This trouble is not serious in grafting nor when current-season scions are used for later-summer budding.

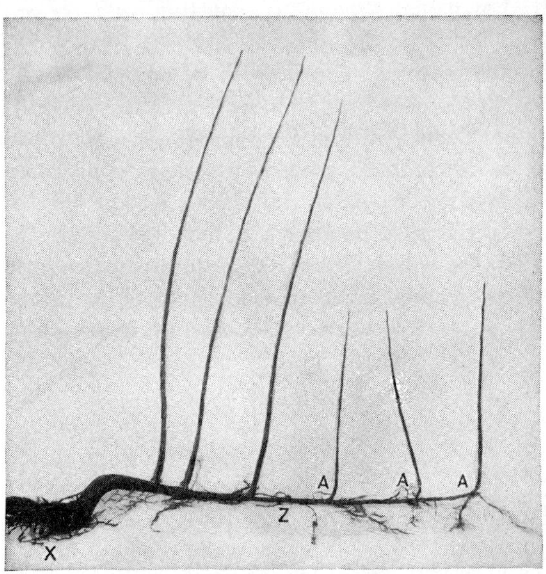

FIG. 125. Trench-layered Paradox walnut showing roots developed during first year. The root system of the original tree is shown at X, the horizontal main stem is indicated by Z, and stems that developed from lateral buds and later formed roots are shown at A. (*Courtesy of E. F. Serr, University of California, Davis.*)

One-year-old walnut seedlings may also be whip-grafted slightly below the ground line during the dormant season. The cleft and inlay grafts are the methods used most commonly in top-working walnut trees.

Propagating Pecans. A good pecan tree for starting an orchard should have a reasonably straight trunk, a height of 5 feet or more, and a well-developed root system.

Producing Nursery Trees. Though several different species can be used as rootstocks for pecans, only the pecan (*Hicoria pecan*) is used in producing nursery trees. Standard varieties cannot be reproduced true to type from seed, but certain selected seedlings and named varieties produce rootstocks of exceptional vigor. The Riverside is popular in Oklahoma, Texas, and New Mexico, and the Stuart is planted most commonly in the southeastern states for the production of rootstocks.

Pecan rootstocks are grown only from seed, which may be planted in nursery rows in the fall; stratified over winter, and planted in the spring; or planted directly in the nursery in the spring after having been presoaked in water for 3 or 4 days. The relatively slow rate of germination is due to the shell, which inhibits the ready absorption of moisture. It is known that the pecan does not have a rest period, since nuts frequently sprout in the hulls before harvest, and they will continue to germinate, if planted under favorable conditions, as long as they remain viable. They lose their viability if they become rancid, and to prevent this they are stored at about 32°F., especially if they are to be kept 1 year or longer before planting.

The nuts are planted about 8 inches apart and covered about 4 inches deep. As soon as the seedlings emerge, they should be cultivated to control weeds and grass and to prevent sunburn at the ground line. Seedlings will normally be large enough to bud by the middle or latter part of the second growing season. The patch bud, or some slight modification of it, is the method used, and buds are forced the following spring. Sometimes the budding is delayed until spring of the third growing season, and the buds are forced shortly afterward. A common method of forcing is to peel the bark from the stock above the inserted bud, leaving the stem for staking the bud to produce a straight trunk. The budded tops are allowed to develop one or two years before the trees are transplanted. There is a limited demand for large trees, and a correspondingly longer time is required to produce these.

Instead of the patch bud, some nurserymen use the whip graft. The one-year-old seedlings are whip-grafted in the nursery in the late winter, and the grafted trees attain suitable size for transplanting usually after two seasons.

Top-working. Frequently it is desirable to top-work pecan trees. This can be accomplished for small trees by patch budding, as used for producing nursery trees. Trees that are up to

6 or 8 inches in diameter can be easily patch-budded by inserting the buds in several limbs that are up to about 3 inches in diameter. The rough outer bark must be pared down to the thickness of the bark of the patch bud for best results. This system is known as *direct budding*. The inlay graft is also a suitable method to use in top-working trees of this size.

Fig. 126. Pecan tree, 6 inches in diameter, top-worked by inlay grafting.

The system commonly employed in top-working very large pecan trees—those up to 2 or 3 feet in diameter—is to dehorn the main branches during the dormant season and to bud the young shoots that develop on the stubs. A very large number of shoots will develop on each stub, and these should be thinned to 5 or 6 per stub, early in the growing season, to encourage vigorous growth. These will be large enough to be patch-budded in late summer, and the buds are forced the following spring

The aftercare of the tree is directed toward encouraging the growth of the inserted buds and to the removal of native sprouts that develop below the buds until their growth is finally restricted by the shade of the new top.

Native hickory trees (*H. alba*) are sometimes top-worked to pecans. The two are not perfectly compatible, and this often-

FIG. 127. Large top-worked pecan tree, showing budded top that has grown one season.

times results in overgrowth of the pecan at the union. The combination, however, is a fairly satisfactory one and the productivity of the trees seems to be determined by the soil in which the hickory trees grow. Those in coarse sandy soil, where moisture is oftentimes insufficient, do not support a pecan top adequately, but in deep alluvial soil with sufficient moisture this difficulty is overcome.

The bitter pecan, or pignut (*H. glabra*), which grows readily in poorly drained locations in extensive areas in the South, is frequently top-worked to pecans. It apparently is a congenial rootstock in its native habitat, and good production is obtained from standard varieties top-worked on it. Poor results have been obtained with this rootstock in highly alkaline soils.

Filberts. Several different methods may be followed in the propagation of filberts. Occasionally a tree grown from seed of a standard variety may have some merit, but the chances are it will be inferior to the parent tree. The filbert produces suckers freely from about the base of the plant. If soil is mounded up around the base of these shoots, they will develop good root systems in time and can be successfully transplanted. Filberts may also be propagated from cuttings. If the cuttings are taken from portions of suckers below the ground, they may root with a measure of success; those that are taken from above-ground parts root very poorly.

Filberts may be propagated readily by continuous layerage. Suckers from around the base are bent down to the ground and pegged in place in early spring. New shoots begin to develop from lateral buds, and as these grow, they are covered by degrees until ultimately 6 to 8 inches of soil is added. These shoots will normally develop a good root system within one season and may be separated for orchard planting during the early winter. Where only a few plants are needed they may be grown from suckers that are tip-layered. Layered shoots from suckers produce plants of the variety represented by the root system of the original plant and so do not reproduce varieties represented by the budded or grafted top.

Budding and grafting are sometimes used in propagating named varieties of filberts. Seedlings of commercial varieties and of the Turkish filbert are grown and whip-grafted to standard varieties. They may also be budded, the T bud being the method used. Large trees may be top-worked by using the cleft graft.

Tung Nut. Trees used in the first commercial tung plantings consisted largely of seedling stock. The variability of seedling plants can be avoided by budding or grafting standard varieties on seedling rootstocks. The patch and ring bud are the two methods that are used most commonly. Seedling trees from seeds planted in February are budded during late summer or early fall. The inserted buds unite but remain dormant over winter. They begin growth the following spring, when they are forced by lopping the tops 5 or 6 inches above the bud. Blind buds are rather common on tung wood, and wood for budding should be selected with some care in order to avoid them.

Persimmon. Named varieties of native and Japanese persimmons are propagated largely by budding or grafting upon seedling rootstocks. The persimmon is considered fairly difficult to propagate.

Rootstocks. In the southern states the native persimmon (*Diospyros virginiana*) is used as a rootstock. In that area it is better than other rootstocks but still not entirely satisfactory. The trees are somewhat difficult to transplant, owing to a long taproot with few lateral roots. Japanese varieties on this rootstock are usually dwarfed by it, and the life of the tree is often 10 years or less.

Seedlings of the Oriental persimmon (*D. kaki*) and the Lotus persimmon (*D. lotus*) are used almost exclusively as rootstocks on the Pacific coast, but named varieties vary in their adaptability to them. Briefly, the Oriental seems suitable for most varieties; it makes a good bud or graft union and is somewhat resistant to crown gall. Like the native American persimmon, it has a long taproot, with few fibrous laterals, and oftentimes responds poorly when transplanted. The Lotus persimmon is the most vigorous rootstock used; it produces a fibrous root system and is considered drought-resistant; and it becomes reestablished readily when transplanted. Unfortunately it is quite susceptible to crown gall and is not satisfactory for certain leading varieties.

Persimmon seeds germinate slowly, because of the slow rate of water absorption. Satisfactory germination can usually be obtained by stratifying the seed and keeping them at low temperature from the time of harvest until they are planted, or by soaking the seed 2 or 3 days before planting. Seeds are usually planted directly into nursery rows. The young seedlings are subject to sunburn, and this can be prevented to some extent by cultivation which will keep loose soil around the bases of the plants.

Budding and Grafting. Seedlings that have grown one season are usually large enough to whip-graft. This is done during the latter part of the dormant season. The trees are grafted in the nursery; and grafts are set slightly below the ground level, sealed with wax, and covered with moist soil. Flipped T buds and patch buds are also used successfully at various seasons of the year, particularly when current-season scions are available as a source of buds. The cleft and inlay grafts are methods adapted to the top-working of trees.

Strawberries. The strawberry plant produces runners freely, which form "rosettes" and take root usually at every second node, forming natural layers. These are the sources of plants for starting a new plantation. Each strawberry plant started in the spring and grown under favorable conditions will readily produce up to 25 layers by fall. Growers in Florida, by a system of separation and replanting during the growing season, harvest from 20,000 to 50,000 layers from an initial spring planting of 1,000 plants. The time for replanting the rooted layers varies with the locality. Where extreme cold occurs during winter, plants are set in early fall so that they will become well established before winter, or planting is delayed until early spring. Where mild winters prevail, the plants are set in fall or early winter for the production of fruit the following spring or early summer. The yellows disease is more prevalent in some sections during the winter, and late-spring planting is a means of avoiding it. The ground should be thoroughly prepared and the rosettes planted so that the crown will be level with the surface of the soil. Those that are set too deep or too shallow recover slowly and make poor growth. In hot and dry climates, better plants can be grown if the strawberry propagating bed is shaded and mulched.

Everbearing strawberries produce few runners, since the buds that normally grow into them form flower buds instead. This characteristic makes it necessary to propagate everbearing varieties by means other than rosettes. Fortunately, plants may be divided by breaking or cutting the crowns apart. Only the large crowns and those that have 10 to 15 roots are satisfactory for planting.

The Bramble Fruits. In general the bramble fruits are given relatively close spacing in the field. Consequently, large numbers of plants are required for an acre, and it is important that they be produced economically. Fortunately the different kinds of brambles are easy to propagate—by one method or another.

Blackberry and Dewberry. Suckers from old plants are a source of new blackberry and dewberry plants. In using these, however, there is always the possibility of getting seedling plants mixed with them, and seedlings are not likely to be desirable plants.

Standard varieties may be readily propagated on a commercial scale from root cuttings. Roots the size of a lead pencil or larger are suitable. These are usually planted in nursery rows in late winter and given a spacing of 6 or 8 inches. They may be cut about 2 inches long, dropped into a furrow, and covered with 3 or 4 inches of soil. They may also be cut about 5 inches long and planted in a vertical position with only the tip of the cutting exposed, in which case it is important that the portion of the root that was nearest the base of the mother plant be placed up. In making the cuttings, the top end may be designated by cutting it squarely across, and the bottom end by cutting it at an angle. Root cuttings grow readily and produce large strong plants in one season. Most varieties of blackberries and dewberries can also be propagated from hardwood and softwood cuttings, planted at appropriate seasons.

Blackberry and dewberry plants may be grown also from trench layers. Vines are covered their entire length in late winter or early spring. New shoots develop from lateral buds, and these form roots on the underground portion. The layers may be separated and transplanted late in the same spring in which they were started, or they may be allowed to grow an entire season before being moved. Trailing varieties produce natural tip and simple layers.

Raspberry. Red raspberries produce suckers, or shoots, which arise in great numbers from the root system after cultivation has ceased. Those that reach a height of 8 or 10 inches make excellent plants for starting a new plantation. They should be lifted with a spade with as much of the root system as possible and replanted in the new location.

Black raspberries and most purple varieties are propagated by tip layers. In many cases, tips of long canes are accidentally covered with soil and natural layers are formed. When many plants are needed, the necessary number of tips should be covered in late summer. With favorable soil moisture, a well-rooted plant will be formed by the end of the growing season. At planting time the following spring, the parent lateral is cut off a short distance above the ground and the new plant dug out with a good portion of the root system for replanting.

The Bush Fruits. The blueberry, gooseberry, currant, and cranberry are the important bush fruits grown in the United States. They are spaced close together in permanent plantings, and relatively large numbers are required for an area.

Blueberry. Some difficulty is experienced in propagating the blueberry and this has retarded the cultivation and improvement of this native fruit. Named varieties can be reproduced from rooted cuttings, by layering, and by budding and grafting.

Most plants are grown from hardwood cuttings; a few from softwood cuttings. Budding and grafting are seldom used except as a step in the rapid multiplication of new varieties. The transfer of buds to vigorous new stems causes rapid growth of the new variety, and these shoots can then be used for making cuttings. Best results in the rooting of blueberry cuttings have been obtained by planting them in a box frame. It has a false floor of hardware cloth to retain the rooting media and to ensure good aeration. The cuttings are made about 4 inches long from one-year-old shoots and planted during late winter or early spring. Peat is a suitable rooting medium. Shading and careful watering are necessary to maintain proper humidity, without encouraging disease. Cuttings usually require 6 to 8 weeks to develop roots, and after this they are gradually hardened in preparation for transplanting.

Gooseberry. This fruit is easily propagated by layers and cuttings. Common varieties are ordinarily started from mound layers. Stock plants from which layers are to be secured should be pruned back heavily before they begin growth in the spring. This stimulates the growth of several shoots from the base of the plant. In midsummer, earth is mounded about the plant so as to cover the lower portion of the new shoots to a depth of 4 or 5 inches. By fall the shoots will have rooted. During the winter the mound is removed, and those cuttings with well-developed roots are cut off and set in a nursery row where they are grown for 1 or 2 years before being finally planted in the field.

Gooseberries may also be grown from cuttings; some varieties root readily and others very poorly. Cuttings about 6 or 8 inches long are made from vigorous-growing shoots during the fall or winter. These are either stored in a cool place for spring planting or set directly into nursery rows. They are spaced from 4 to 5 inches apart and set at a depth that permits only two buds to extend above the ground. New plants suitable for field planting are produced from cuttings in one or two seasons.

Currant. New currant plants are grown almost entirely from cuttings made from vigorous shoots that have completed one season of growth. They are cut about 8 inches long and planted as soon as they are made, or stored in a cool place during the fall or winter for spring planting. In either case they are set in rows with from 4- to 6-inch intervals between plants and at a depth that permits only one or two buds to extend above the ground. Better rooting is obtained if the soil in which the cuttings are placed is well-drained and is sufficiently loose and porous to permit good aeration. Some cuttings will produce plants suitable for field planting in one season; others require two seasons.

Cranberry. The cranberry is propagated commercially from cuttings, which are planted directly in place without previous rooting. Most cranberry plantings are made on acid peat soils. A surface layer of sand, 4 or 5 inches deep, is added to the peat before the area is planted. The cuttings are pressed into the sand so that the lower portion is in contact with the peat. It is customary to give cuttings a spacing of from 10 to 12 inches each way.

Citrus. In the few states where citrus is grown on a commercial basis the general method of propagation follows a rather definite procedure. The difference in soil and other environmental factors makes the choice of stock a consideration of primary importance. Within recent years, the relationship between rootstock and disease, particularly the virus causing "quick decline" or "tristeza," has resulted in a careful reexamination of the entire question of rootstocks.

Rootstocks. Sour orange was for many years the most widely used rootstock for citrus, not only in the commercial areas of California, Florida, and Texas, but also in many other parts of the world. It is resistant to foot rot and gum disease of various kinds, makes a good compatible union with orange and grapefruit, resulting in a tree of moderate size and vigor and of outstanding productivity. It has been found, however, that citrus trees, especially orange, on sour orange, manifest extreme symptoms of quick decline, while trees of the same variety on several other rootstocks apparently survive in good condition. Since this virus disease has now been found in most of the commercial citrus areas of the United States, the sour orange has been partially replaced in many areas and is subject to suspicion in most others. There are, nevertheless, many old bearing citrus trees now growing on sour-orange rootstock, so that the problem will be in existence for years to come.

Fig. 128. Citrus-seedling rootstocks. Left to right, Rusk citrange, Cleopatra mandarin, rough lemon, sour orange, sweet orange, Tresca pomelo.

Sweet orange has been used extensively in California as a rootstock for lemon and sweet orange. It is entirely compatible and makes a good tree, although the fruit tends to be small. It is quite susceptible to brown-rot gummosis and is not adapted to heavy or poorly drained soils.

Rough lemon has been used extensively as a rootstock for citrus on light soils in Florida. It has the ability to make a much larger tree on such soils than does sour orange. The trees are also very productive, but the fruit tends to be somewhat less desirable in texture and quality than that on either sweet or sour root. It has been used to a very limited extent in Texas, and in California it has been found more susceptible to cold injury than other common rootstocks.

Cleopatra mandarin is probably the most promising of the newer citrus rootstocks. It is one of the most tolerant to the quick-decline virus, and is also resistant to other diseases. In a recent comparison in Texas it has been found to give a slightly higher percentage of successful bud unions than sour orange, although difficulty in budding has been a criticism of this stock. It appears to be compatible with most citrus varieties, and is replacing sour orange to some extent.

Several new hybrids, principally citrange and tangelo, have shown considerable promise as rootstocks. Troyer citrange is vigorous and resistant to cold in California, while in Texas the Rusk, Carrizo, and Uvalde citranges have given good results. The Sampson tangelo is apparently tolerant to quick decline with lemons in California, but orange on this stock is susceptible to the virus.

The trifoliate orange is the standard stock for the entire Gulf coast, outside the areas of commercial production. It is widely used along the coast from Corpus Christi, Texas, to northern Florida. It is universally used as a stock for the Satsuma orange, which does not succeed on sour orange; trifoliata is commonly used, also, for kumquat and Ponderosa lemon and for any of the grapefruit and round oranges in the coastal region mentioned. Trifoliata is a deciduous species and is hardy to cold. It is frequently stated that it imparts this cold resistance to the top budded on it, but there is little evidence to support this belief. It is not tolerant to saline soils.

Many other rootstocks are now on trial, in a comprehensive program of testing for resistance to quick decline, adaptability to saline or alkaline soils, and also for compatibility with the important commercial varieties of citrus.

Citrus seed is nearly always planted in a special seedbed, where better germination may be obtained and better attention given to the young plants. Most of them will need protection against scab. The young plants at the end of a flush of growth will be 8 to 12 inches high. They are dug with a spading fork; all weak, crooked, or otherwise undesirable plants are culled out; and the good plants set in the nursery row about 12 inches apart, the rows being 5 to 6 feet apart. They are allowed to attain a diameter of ¼ to ⅜ inch before they are budded.

Budding. Commercial citrus trees are propagated by budding, which is not considered difficult. Budwood should be taken from trees of known variety and preferably from those certified free of virus diseases. The best budwood is from the next to the last flush of growth, which has become rounded instead of angular, as the new wood is in its early growth. The shield or T bud is used, either upright or inverted. Budding can be done at almost any season when the bark will slip well. There is an advantage

FIG. 129. Budded citrus trees, staked for training.

in budding in early spring, as there is less danger of losing valuable buds from cold; and also the buds set at this time will make good trees in one season. Dormant budding is also used by many nurserymen. The buds are cut from the bud stick with a sliver of wood and not slipped as in the peach. Rubber budding strips are used extensively for tying, and strips of waxed cloth are also used to a limited extent.

Training of Tops. As the buds are forced and the old tops removed, the training of the young shoot receives careful attention. It is tied at close intervals to a 1- by 1-inch stake, and a straight shoot is secured. The common practice is to allow this shoot to reach a height of about 36 inches, then to head it back to 30 inches, and finally to force out a few branches to develop the permanent framework. Citrus nursery trees are usually sold as two-year trees, with tops already developed. This is one reason, aside from the fact that they are evergreens, why citrus trees are usually sold as balled-and-burlapped (B&B) and seldom bare-rooted.

Own-rooted Plants. The production of the various species of citrus on their own roots is desirable for certain purposes. Uniform stocks might be produced, not only for experimental use but for commercial orchards as well. Some species and varieties,

Fig. 130. Citrus trees, balled and burlapped, ready to plant in the orchard.

for which no stock of satisfactory performance has been found, are best propagated as own-rooted plants.

Cuttings of most of the species of citrus, and some related genera, have been rooted in California. Stem cuttings 4 to 6 inches long with about seven nodes have given the best results. The lower leaves were removed, but the three or four upper ones were left. Rooting percentages of 75 per cent and upward were obtained. The Meyer lemon is produced commercially by this method.

Fig. This fruit has always been found easy to propagate, and nursery trees can be produced on a large scale at very little expense.

Cuttings. The common method of propagating figs is by means of hardwood stem cuttings, which are usually made from wood of the previous season. Wood with comparatively short internodes is preferable to that in which they are extremely long. The cutting should have a minimum of two nodes, but those of the proper length, approximately 8 inches, will have three or more nodes. The terminal portions of the shoots may be used exclusively for cuttings, but in seasons of moderate injury from cold these tips may be killed, when the rest of the wood is unharmed. A satisfactory practice is to remove entire shoots of one-year wood and cut them into proper lengths.

The cuttings of figs should be made in the latter part of the dormant season, usually in January or early February. They are set directly in the nursery row in beds where the soil has been well prepared and where adequate provision has been made for drainage. If the soil permits, the row may be marked off and the cuttings pushed down into the loose earth. Another method is to open the bed with a plow and set the cuttings with enough soil pressed against them to hold them in place. They can also be set satisfactorily with a mechanical transplanter. The row is then covered by the plow. In either case the cuttings should be set deep in the ground so that only one bud is exposed. In a dry season one irrigation in early spring will be of great value in promoting growth of the cuttings. Under favorable conditions, good rooting occurs resulting oftentimes in a near-perfect stand. Close setting of the cuttings, about 8 inches apart

Fig. 131. Fig tree grown from cutting, showing one year's growth.

in the row, provides a satisfactory stand. These cuttings will make good trees by fall and frequently produce fruit the same season.

Suckers are sometimes used for starting new trees, but diseases, insects, or other organisms present on the parent tree are likely to be transferred to the new location.

Budding. Top-working of figs may be done with little difficulty. It has been shown that young fig trees may be budded easily with the T bud. The same method is successful also on large trees, in wood one to three years old, provided budwood of large diameter is available. This method has been used successfully at various times during the growing season whenever the bark was slipping. Patch buds have also been used with good results. It is advisable in either case to place the buds early in the season, using storage budwood, and force them out, in order to get the maximum amount of growth.

Grafting. The cleft graft and bark graft are used more commonly in top-working of figs. Cleft grafting is successful on branches and trunks 3 or 4 inches in diameter. The work is done in late winter or early spring, before growth starts. Bark grafting is done in March or April after the bark has begun to slip. This method, and also the preceding one, are carried out in accordance with general instructions previously mentioned. It is advisable in either case after the wax has been applied to cover the stubs and scions with whitewash to prevent sunburn. The use of paper sacks as coverings is also recommended to prevent drying of the scions. All grafts, but more specifically bark grafts, require staking or bracing to prevent their being damaged by wind.

Cultivated varieties are very susceptible to root-knot nematode attack. The Palmata fig (*Ficus palmata*) is resistant to nematodes and is being tried as a rootstock for figs to be planted in infested soil. It can be grown readily from cuttings. Unfortunately the species is more tender to cold than most varieties of figs.

Seed. In the case of varieties that produce viable seed this method of propagation can be followed. Its use, however, is restricted almost entirely to breeding work.

Avocado. Avocadoes are grown commercially in California and Florida, and to a limited extent in Texas. Standard varieties are propagated entirely by budding and grafting them on seedling rootstocks.

Rootstocks. As with many other fruits, the question of rootstocks is highly important in the case of the avocado. In California, seedlings of the Mexican race are used commonly because of their greater resistance to cold. Seedlings of some of the hybrids of Mexican and Guatemalan races are also used to a limited extent. In Florida, the West Indian seedlings are used exclusively, because of their natural adaptation to the area and because cold hardiness is not as important as in California. In Texas, West Indian rootstocks are also used in preference to the Mexican, because of their tolerance to saline soils. Mexican seedlings and also the commercial varieties grafted on them show considerable tipburn, which is attributable to salinity.

The seeds should be planted as soon as possible after removal from the fruit; but if necessary, they can be stored in a cool place in materials such as peat moss, sand, or sawdust, with emphasis on the fact that they should not be allowed to dry out. The seeds are nearly always sprouted in a seedbed before planting for propagation. In California, they are lined out in the nursery for budding; while in Florida and Texas they are placed in No. 10 cans and grafted before being planted. As a preliminary treatment before planting, the seed should be cut transversely at the apical or pointed end, to remove about $\frac{1}{4}$ to $\frac{1}{2}$ inch of the top. This cut removes the interlocking tips of cotyledons and permits the young growing tip to emerge more readily.

Budding and Grafting. In California, the avocado is successfully T-budded in the nursery, with the buds being well wrapped with rubber strips or strips of cloth or pliofilm. Great care is necessary in the selection of budwood, which is best taken from the terminal portions of well-matured flushes of growth, with fully matured leaves. Poor selection of budwood has resulted in many unsatisfactory experiences in budding. The same method appears to be very satisfactory in Mexico and many countries in the tropics. In Texas, the T bud has not been satisfactory, and part of the difficulty, at least, is due to the inability to obtain good budwood. In this area, the most successful method is a cleft graft on a young seedling growing in a can and with wood still in rather immature condition. The grafts succeed best if stock and scion are nearly the same in diameter. The graft is wrapped with nursery tape and also a rubber strip to maintain an even pressure. The completed grafts are placed in chambers made of polyethelene material, to maintain a high humidity. The graft may be expected to unite and begin growth in 30 to 45 days after they are made.

Fig. 132. Avocado grafted by approach—now on seedling rootstock.

This same method has been described as used in Florida, but the side graft is used more commonly. On these young stocks the T bud was formerly used to a considerable extent, but it was found that some varieties were difficult to bud and that much wood unsuitable for budding could be used in grafting. At present, the side graft and the cleft graft on young trees in individual containers are the most important commercial methods in Florida.

Top-working of avocado has been done with considerable success in Florida by means of a cleft graft, in which a saw is used to make a cut 3 to 5 inches deep. The grafts are waxed and then the entire stub and graft are covered with a wide strip of paper tied at the base and open at the top so that it can be filled with moss or other similar material, which can be watered at intervals. Polyethylene bags may be used where the sun is not sufficiently intense to cause burning.

Jujube. Named varieties of jujube are easily propagated by grafting onto seedling rootstocks. Seeds are stratified and kept cool over winter. They are planted in the spring and will develop seedlings large enough for whip grafting at the beginning of the second year. The grafts are inserted at the ground line or slightly below. Some varieties produce suckers freely, and these are useful in developing new plants.

Pomegranate. The pomegranate is propagated readily from hardwood cuttings. They are normally planted directly into nursery rows in well-drained sandy loam soil. Under favorable conditions they make sufficient growth to be transplanted after one season's growth.

Olives. Olives are propagated principally by softwood and hardwood cuttings. The Sevillano, one of the leading commercial varieties, is commonly grafted onto seedling rootstocks because it grows poorly from cuttings that are planted in the ordinary manner. Leafy softwood cuttings of this variety will, however, form roots and produce good plants if treated with a hormone and grown under mist humidification.

Guava. The guava is propagated from seed since it commonly comes true to type, and also by whip and bark grafting named varieties onto seedling rootstock at the crown. They do not grow well from cuttings, but can be grown from layers.

Lychee. The lychee, a subtropical fruit, is propagated entirely by air layering. Some success will be attained with layers made during the different seasons of the year, but best results are obtained with those made as soon as danger of frost is past in the early spring. Limbs about ½ inch in diameter are selected.

The usual techniques of girdling, hormone treatments, and the use of moist sphagnum and plastic sheets are employed. Normally 3 or 4 months are required for rooting, and the rooted layers are then removed and planted in No. 10 cans or tar-paper pots. After potting, the plants should be kept protected from full-sun and wind exposure until they have become fully established and started growing.

Papaya. Seedlings from seed that have been produced under controlled pollination are used most commonly in producing the papaya. The papaya is normally dioecious and pollination can fortunately be controlled easily. Vegetative propagation of varieties is practiced by topping old plants to produce side shoots from which heel cuttings are made. These shoots are partly defoliated when removed and are provided with bottom heat in the cutting bed to encourage rooting.

Dates. Offshoots are used for propagating named varieties of dates. Seed propagation is not satisfactory because about half of the seedlings are staminate plants which are unfruitful, and fruit of the pistillate plants that are produced is nearly always inferior to that of the mother plant. Offshoots that arise from axillary buds at or near the ground line are encouraged to develop roots, by mounding soil around the base of the date palm, and these are later transplanted to permanent locations.

Rose. The rose has been a popular flower since the earliest periods of history. Its great range of forms and colors adapt it to a wide variety of uses. Methods by which roses are propagated vary with the species and the geographic locality; seeds, cuttings, layers, suckers, buds, and grafts are used.

Seeds. New varieties of roses originate largely as selections from seedlings. Occasionally, one orginates as a sport. Some true species, as *Rosa multiflora,* may be economically reproduced from seeds. The rose fruits, or "hips," should be collected as soon as ripe and the seed removed. From this time until planted the following spring it is important that the seed be kept continuously moist. Immediate stratification of the seed in sand or some other moist material is recommended. It is important that they never be allowed to become dry. Experiments indicate that the best temperature during stratification is about 41°F. The seeds of some species apparently have a longer rest period than others. Stratification at 41°F. for 270 days is recommended for those of the dog brier (*Rosa canina*). Ninety days is sufficient for the Pasture rose (*R. humilis*) and 50 for the multiflora (*R. multiflora*). Stratification serves to keep seed from drying out and thus preserves their viability, and cold temperature breaks the rest period. Seed stratified in late summer or early fall are commonly planted the following spring in either nursery rows or special seedbeds. The seedlings may be transplanted to their permanent location at the end of the first growing season.

Cuttings. In general, roses grow readily from cuttings, and no other method is followed so extensively. Cuttings are used in the multiplication of named varieties and also for the starting of stocks for budding or grafting.

Fig. 133. Rose cuttings are planted in the rows by hand, after which the soil is pressed closely about the cuttings. (*Courtesy of A. F. Watkins, Dixie Rose Nursery, Tyler, Tex.*)

Rosewood in various stages of maturity may be used for cuttings. A comparatively simple method of increasing plants of some species is by means of *dormant hardwood cuttings* made in late fall or winter. The cuttings, usually made from canes about the size of a lead pencil, are cut into lengths from 6 to 8 inches. A rule in making hardwood cuttings is to cut immediately below a node at the base and immediately above a node at the terminal. As a matter of expediency, this is usually disregarded in making rose cuttings. The canes are tied into bundles, and cut to length with a bandsaw or special cutter without regard to the position of basal and terminal nodes on each cutting. Removing all buds except the one at the tip node, a practice known as disbudding, tends to discourage the development of water sprouts later in the growth of the plant. It is important that the bottom of the cutting is placed downward and that it is not planted in an inverted position. Uncertainty as to which is the bottom end may be dispelled by examination of the leaf scar, which may be seen ordinarily on the stem beneath each bud. Most hardwood cuttings for the production of rootstocks are planted directly into the nursery rows in the field. This is usually done during the latter part of January or February. The rows are usually 6 feet apart, and cuttings are spaced about 8 inches apart in the row. Better rooting will be obtained if the cuttings are planted in a well-drained sandy or sandy loam soil. Drainage may be facilitated by planting on beds from 4 to 6 inches high. By the time top growth starts in the spring, the cuttings will have formed roots. The plants that result develop to a size suitable for budding by June of the same year. At that time they may be budded, but more commonly budding is delayed until August. These phases of propagation will be discussed in a paragraph that follows. Cuttings that are not to be budded, for example, those named varieties that are to produce "own-rooted" plants, are allowed to grow in the nursery rows one or two seasons.

Fruit jars are sometimes inverted over hardwood rose cuttings that are planted about the house or yard. The beneficial effect of these is probably due to the maintenance of high humidity about the cutting. Commercial growers accomplish the same results by growing the cuttings under a glass sash or cover and by watering them at frequent intervals.

In some sections the rooting of hardwood cuttings is not entirely successful. Another practice followed to some extent is to grow the rose plants from *softwood cuttings* made from material during its current or first season of growth. Such material is soft, immature, and succulent, and cuttings made from it require more care and attention in the propagation bed. The cuttings, 6 to 7 inches long with only the basal leaves removed, are planted upright about 3 inches deep. They are usually set in a special propagation bed with sand as the rooting medium and are shaded with cheesecloth until rooted. Slatted frames offer the same kind of protection. Frequent waterings are necessary to keep the leaves from wilting and dropping. These cuttings root commonly in from 10 to 14 days—a shorter period than required for hardwood cuttings. As new growth starts, indicating root formation, the shade is gradually removed, first for only a part of each day and finally for the entire day. Softwood cuttings should be made as early in a season as the wood becomes sufficiently mature. In the South, April or early May is not too early; in the North they are made the latter part of July and August. If made early, they form a better root system and have a longer period during which to mature. Cuttings may be taken from any part of a current-season shoot. Those taken from the tips of growing canes root more quickly than others taken from the base or middle portion. The difference, however, is very slight and, since plants started from the older parts of canes often mature more properly in the fall, many propagators prefer to use the older material. One serious objection to softwood cuttings is that they cannot be disbudded easily. At the end of the first season, plants from softwood cuttings may be used as lining-out stock; in the case of named varieties the plants may be moved to their permanent location.

Stocks. Most species of roses are congenial when intergrafted or interbudded. Theoretically, then, almost any species could be used as a stock. Different species, however, vary widely in the many characteristics that determine their merits for stock purposes. Some species, for example, root readily from cuttings; others, very poorly. Still others, although they grow readily from cuttings, form weak and poorly developed root systems. Likewise there are differences in such characteristics as suckering, tolerance of or resistance to nematodes, and the presence or absence of thorns. In actual practice, only a few species are used for stock purposes. Thornless strains of *Rosa multiflora* are the most popular stocks. Cuttings of it can be handled more easily, and the resulting plants can be budded more readily than is the case with species that have thorns. The species roots readily, forms a good root system, and is decidedly resistant to nematodes. It seldom forms suckers. The dog brier (*R. canina*) is an excellent rose stock but has very heavy thorns that make it difficult to handle. It has a tendency to form suckers in some regions. *Rosa manetti* and *R. odorata* are the principal stocks used for greenhouse roses. *Rosa rugosa* is used as a stock for "tree roses."

Grafting. Grafting was long a standard practice in the propagation of greenhouse roses. One-year-old rootstocks grown from cuttings may be whip-grafted with the beginning of the second growing season. The stocks may be dug and the work done indoors during the winter, or they may be grafted in the field in early spring. In either case plants of commercial planting size are produced in one season.

Grafting is one method of propagating rose plants for forcing purposes. The plants are grafted in midwinter and carried along

Fig. 134. Dormant rose buds forced in early spring, showing growth up to June 1. (*Courtesy of A. F. Watkins, Dixie Rose Nursery, Tyler, Tex.*)

in small pots until sold by the propagator in the late winter or early spring. The plants may be held for a time in pots by the grower, but they are usually planted in the greenhouse benches by midsummer for production of cut flowers the following winter. In sections where winters are mild, rose plants may not become completely dormant, in which case it is difficult to obtain scions suitable for grafting. The susceptibility of grafted plants to the attack of crown gall discourages the practice of grafting roses.

Budding. The T bud is the method followed quite universally in budding rootstocks grown from seed or cuttings. Buds are taken from bud sticks of the current season's growth. It is important that the scions be cut early in the morning, that the leaves be removed by clipping leaf petioles ⅓ inch out from the bud, and that the bud sticks be kept in a cool, moist place until used. A fresh supply of budwood should be cut each day whenever it is practicable to do so. If insulated in moist moss or other material and kept at a temperature of 32 to 40°F. they may be safely used over a period of 4 or 5 days or longer.

Hardwood cuttings planted in the field in late winter or one-year plants lined out in early spring make sufficient growth in warmer parts of the United States to be budded in early summer. Budding that is done in June is referred to as *June budding,* and plants that are produced thereby are known as *June buds.* June-budded rose plants may make from 18 to 24 inches of growth before the end of the first season. They are not likely, however, to make sufficient growth to be classed as No. 1 plants. An advantage of June budding is that the grower is able to produce a rosebush suitable for planting within one growing season.

Instead of budding the rose stock in June, the operation may be delayed until late summer. Buds are inserted at or slightly below the ground line. To do this it is necessary to remove a surface layer of soil from around the base of the plant. The buds are placed in the smooth part of the cutting that was disbudded at planting time and not in the new growth from the cutting. Raffia and rubber bands are commonly used for tying. Twine also may be used. No wax or similar material for excluding air is necessary. The buds unite with the stock but are not forced into growth until the following spring. Since they remain dormant over winter, the method is known as *dormant budding.* In winter or very early spring the native top is cut off, and the inserted bud begins growth in early spring as a result of the stimulus. When it has made 6 or 8 inches' growth, it is pruned to encourage branching. Under favorable growing conditions it develops into a large, vigorous plant by the end of the growing season. The stock of such a plant has two seasons in which to make its growth, while the top has only one. If as a result of drought or other unfavorable conditions the rose stocks cannot be budded in late summer, the budding may be done early the following spring. The buds are forced immediately and produce strong rose plants by the end of the growing season.

A common procedure in propagating roses is to plant rose

cuttings and bud them after they have rooted and made some growth. Instead, buds may be inserted in branches of a plant that is to be used later for cuttings. This is known as *cane budding.* As many as 8 or 10 buds spaced 5 to 7 inches apart are inserted in a long cane. The time at which this is done depends on the most favorable season for making the cuttings. The several buds on a cane unite with it, within a period of 2 weeks. Immediately after union, if they are handled as softwood cuttings, and the following winter in case of hardwood cuttings, the canes are cut into as many cuttings as there are living buds.

FIG. 135. Cane budding of multiflora rose.

All native buds are removed. When the cutting has rooted and the shoot begins to develop from the bud, it is best that the shoot be cut back to about three strong buds so that it will make a bushy plant.

Certain species of roses make a very rank growth and produce strong, relatively inflexible canes. With proper pruning, such species can be caused to grow into a single upright stem for 4 or more feet from the ground. When such stocks are budded at a height of 4 or 5 feet above the ground, and the bud develops into a top or bush, the resulting plant is known as a *standard* or *tree* rose. Tree roses are interesting novelties wherever found, and in some sections have value for use in landscaping. Any upright-growing species that has been trained to a single standard can be used as a stock; **R**. *rugosa* is used more commonly than any other. The beauty of the roses will be appreciated more if a variety with a long blooming season is used for a top.

Layers and Suckers. Most varieties of roses can be reproduced from layers. The method may be used for the production of a limited number of plants, but it is seldom used in commercial propagation. Likewise, some varieties produce suckers freely and these may be separated from the parent plant and used as a source of new plants.

Own-rooted, or Budded and Grafted Roses. Rose plants that are allowed to grow to maturity from cuttings are known as own-rooted plants. Those that might be grown to a mature size from seed likewise would be regarded as own-rooted. In general, roses grow readily from cuttings of one kind or another. It would seem, then, that there would be no occasion for growing any kind other than own-rooted plants. Some few varieties, however, do not root well. Others, although they grow from cuttings, form weak root systems that are not adapted to certain soils, or perhaps are not tolerant to pests that are prevalent in some soils. Certain varieties on their own roots are perfectly acceptable for some conditions, while under a different set of conditions they respond very poorly.

Fig. 136. *Rosa rugosa* used as a rootstock for tree roses.

Budded or grafted rose plants consist of two components: the stock and the top that develops from the bud or graft

FIG. 137. Variation in rose plants grown in same nursery row, showing the need for grading standards. (*Courtesy of A. F. Watkins, Dixie Rose Nursery, Tyler, Tex.*)

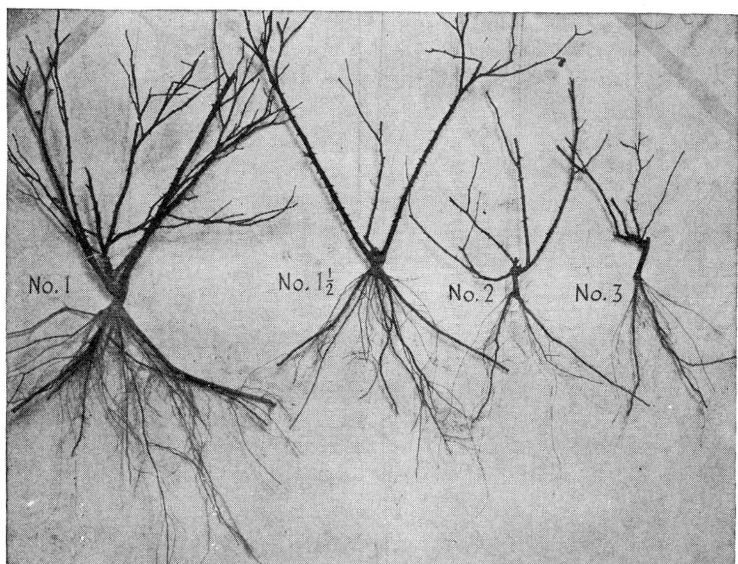

FIG. 138. Single plants showing characteristics of the commercial grades of roses. (*Courtesy of A. F. Watkins, Dixie Rose Nursery, Tyler, Tex.*).

inserted in the stock. The propagator can select a stock and scion each with certain desirable characteristics and by budding or grafting grow a plant that is more useful in many cases than would be possible with own-rooted plants.

QUESTIONS

1. What is the objection to grafting peaches?
2. Name the different rootstocks used for peaches. What are the merits of each?
3. How is the Marianna plum rootstock propagated?
4. How are peach seed treated in order to improve germination?
5. What method of budding is used most in propagating the peach? What kind of budwood is used?
6. Outline the successive steps, with dates of each, in propagating a June-bud peach tree; a one-year-old peach tree.
7. What methods may be used in top-working a large peach tree?
8. What rootstocks are used for plums? Apricots? Cherries? Almonds?
9. How should apple and pear seed be handled in order to ensure good germination?
10. Outline the various methods used in propagating apple trees suitable for orchard planting.
11. Tell how to top-work a large apple tree.
12. How are dwarf-apple trees produced? Of what value are they?
13. What seedling rootstocks are used for pears? What intermediate stocks?
14. Name the various methods of budding and grafting used in propagating the pear.
15. Tell how to produce a pear tree with blight-resistant framework; a dwarf-pear tree.
16. How should pecan seed be treated in order to ensure good germination?
17. What method would be followed in budding a small pecan tree? In budding one 6 inches in diameter? In top-working one 15 inches in diameter?
18. What rootstocks are used for Persian walnuts? For black walnuts?
19. What methods are used in propagating persimmons?
20. List the different ways of propagating grapes.
21. Grape plants that are grown from cuttings are used for what purposes?
22. What are the requirements of a good rootstock for grapes?
23. Outline the different ways by which grape plants suitable for planting in the vineyard are propagated.
24. How are blackberries propagated? Dewberries? Raspberries? Strawberries? Blueberries? Gooseberries?

25. What different rootstocks are used for the various kinds of citrus?

26. Outline the steps in producing nursery trees of citrus.

27. What different methods are used in propagating figs?

28. Outline the different ways of handling hardwood cuttings of rose.

29. What are the requisites of a good rose stock?

30. What is cane budding? How are "tree roses" propagated?

31. How are each of these propagated: Date? Avocado? Filbert? Lychee? Olive?

SUGGESTED REFERENCES

Stone Fruits

Day, L. H.: Rootstocks for Stone Fruits, *Calif. Agr. Expt. Sta. Bull.* 736, 1953.

————: Cherry Rootstocks in California, *Calif. Agr. Expt. Sta. Bull.* 725, 1951.

Hanson, C. J., and E. R. Eggers: Propagation of Fruit Plants, *Calif. Agr. Ext. Serv. Circ.* 96, 1951 rev.

Howe, G. H.: Mazzard and Mahaleb Root Stocks for Cherries, *N.Y. (Geneva) Agr. Expt. Sta. Bull.* 544, 1927.

Hutchins, Lee M.: Nematode-resistant Peach Root Stock of Superior Vigor, *Proc. Am. Soc. Hort. Sci.*, **34:**330–338, 1937.

Day, L. H., and Warren P. Tufts: Nematode-resistant Rootstocks for Deciduous Fruit Trees, *Calif. Agr. Expt. Sta. Circ.* 359, 1944.

Van Alstyne, L. M.: The Plum in New York, *N.Y. (Geneva) Agr Expt. Sta. Circ.* 134, 1932.

Pome Fruits

Argles, G. K.: A Review of the Literature on Stock-scion Incompatibility in Fruit Trees, with Particular Reference to Pome and Stone Fruits, *Imperial Bur. Fruit Production, East Malling, Kent, England, Tech. Bull.* 9, 1937.

Childs, Leroy, and Gordon G. Brown: A Study of Tree Stocks in Relation to Winter Injury and Its Prevention, *Oregon Agr. Expt. Sta. Circ.* 103, 1931.

Day, L. H.: Apple, Quince, and Pear Rootstocks in California, *Calif. Agr. Expt. Sta. Bull.* 700, 1947.

Reimer, F. C.: French Pear Rootstocks, *Oregon Agr. Expt. Sta. Bull.* 485, 1950.

Walnut and Pecan

Bailey, J. E., and J. G. Woodroof: Propagation of Pecans, *Georgia Agr. Expt. Sta. Bull.* 172, 1932.

Brison, F. R.: The Storage and Seasoning of Pecan Bud Wood, *Texas Agr. Expt. Sta. Bull.* 478, 1933.

Serr, E. F., and H. I. Forde: Comparison of Size and Performance of Mature Persian Walnut Trees on Paradox Hybrid and *J. hindsii* Seedling Rootstocks, *Proc. Am. Soc. Hort. Sci. Vol.* 57, 1951.

Sitton, B. G.: Vegetative Propagation of the Black Walnut, *Mich. Agr. Expt. Sta. Tech. Bull.* 119, 1931.

Grape

Husmann, George C.: Testing Phylloxera-resistant Grape Stocks in the Vinifera Regions of the United States, *U.S. Dept. Agr., Tech. Bull.* 146, 1930.

Jacob, H. E.: Grape Growing in California, *Calif. Agr. Ext. Serv. Circ.* 116, 1950.

Winkler, A. J.: Some Factors Influencing the Rooting of Vine Cuttings, *Hilgardia,* **2**:329–349, 1927.

Berries

Colby, A. S., H. W. Anderson, and W. P. Flint: Bramble Fruits, *Illinois Agr. Expt. Sta. Circ.* 427, 1937.

Baker, R. E., and H. M. Butterfield: Commercial Bush Berry Growing in California, *Calif. Agr. Ext. Serv. Circ.* 169, 1951.

Eaton, E. L.: Blueberry Culture and Propagation, *Dominion Dept. Agr. Farmers' Bull.* 120, 1949 rev.

Johnson, Stanley: The Propagation of the Highbush Blueberry, *Mich. Agr. Expt. Sta. Spec. Bull.* 202, 1930.

Waldo, George F., and O. T. McWhorter: Crown Division—A Means of Propagating Everbearing Strawberries, *Oregon Agr. Ext. Serv. Bull.* 488, 1936.

Citrus

Camp, A. F.: Citrus Propagation, *Florida Agr. Ext. Serv. Bull.* 139, 1950 rev.

Hanson, C. J., and E. R. Eggers: Propagation of Fruit Plants, *Calif. Agr. Ext. Serv. Circ.* 96, pp. 50–52, 1951 rev.

Webber, H. J.: Variations in Citrus Seedlings and Their Relation to Rootstock Selection, *Hilgardia,* **7**:1–79, 1932.

Fig

Condit, I. J.: Fig Culture in California, *Calif. Agr. Expt. Sta. Circ.* 77, 1933.

Stansel, R. H., and Wyche, R. H.: Fig Culture in the Gulf Coast Region of Texas, *Texas Agr. Expt. Sta. Bull.* 466, 1932.

Avocado

Haas, A. R. C.: Propagation of the Fuerte Avocado by Means of Leafy-twig Cuttings, pp. 126–130, *Calif. Avocado Assoc. Yearbook,* 1937.

Hanson, C. J., and E. R. Eggers: Propagation of Fruit Plants, *Calif. Agr. Ext. Serv. Circ.* 96, pp. 43–49, 1951 rev.

Hodgson, Robert W.: The California Avocado Industry, *Calif. Agr. Ext. Serv. Circ.* 43, 1934.

Maxwell, Norman P.: Avocado Propagation in the Lower Rio Grande Valley of Texas, *Texas Avocado Soc. Yearbook*, 30–32, 1953.

Rose

Tukey, H. B.: The Propagation of Multiflora Rootstocks for Roses by Soft Wood Cuttings, *N.Y.* (*Geneva*) *Agr. Expt. Sta. Bull.* 508, 1931.

Weinard, F. F., and H. B. Dorner: *Rosa Odorata* as a Grafting Stock for Indoor Roses, *Illinois Agr. Expt. Sta. Bull.* 290, 1927.

—— and S. W. Decker: Summer-budded versus Winter-grafted Roses, *Illinois Agr. Exp. Sta. Bull.* 358, 1930.

—— and ——: Effects of Prolonged Storage on Forcing Qualities of Summer-budded Roses, *Illinois Agr. Expt. Sta. Bull.* 409, 1934.

Miscellaneous Fruits

Chadwick, L. C.: Studies in Plant Propagation, *Cornell Univ. Agr. Expt. Sta. Bull.* 571, 1933.

Camp, A. F., and Harold Mowery: The Cultivated Persimmon in Florida, *Florida Agr. Ext. Serv. Bull.* 124, 1945.

Cobin, Milton: The Lychee in Florida, *Florida Agr. Expt. Sta. Bull.* 471, 1950.

Hansen, C. J., and E. R. Eggers: Propagation of Fruit Plants, *Calif. Agr. Ext. Serv. Circ.* 96, 1951 rev.

Hartmann, H. T.: Leafy Sevillano Olive Cuttings, *Calif. Agr., vol.* 8, no. 5, 1954.

Hitchcock, A. E., and P. W. Zimmerman: Relation of Rooting Response to Age of Tissue at the Base of Greenwood Cuttings, *Contrib. Boyce Thompson Inst.*, 4:85–98, 1932.

Mowery, Harold, L. R. Toy, and H. S. Wolfe: Miscellaneous Tropical and Sub-tropical Florida Fruits, *Florida Agr. Ext. Serv. Bull.* 156, 1953 rev.

Watkins, John B.: Propagation of Ornamental Plants, *Florida Agr. Ext. Serv. Bull.* 150, 1952.

CHAPTER 15

Transplanting

Transplanting consists of moving plants from one place to another with the intention of having them continue their growth in the new location.

The art of transplanting is probably practiced more widely than any other in horticultural work, except that of planting seed. It is important in the growing of flowers, vegetables, and fruits. Many vegetable crops are started in specially prepared seedbeds and later moved to the field. Building sites are quickly made attractive, parks are established, highways are provided with shade, orchard and small-fruit plantations are established, forests are replanted, and flowering plants are rendered more valuable—all by various adaptations of this practice. The distance involved may be small or great, only a few feet or hundreds of miles. Success in either case depends

Fig. 139. Self-propelled two-row transplanter for closely set plants like onion.

partly upon care exercised in the three rather distinct operations of digging, moving to the new location, and replanting. It depends, also, on the kind of plant, the condition of the plant, and upon certain environmental factors, as, for example, humidity and temperature.

When a plant is transplanted, it may resume growth in due

time—either promptly or delayed—or it may die. To survive, the plant must have sufficient reserve-food materials to sustain respiration and to support the initial growth of roots and top. In addition, it must have, or it must develop quickly, roots to take up sufficient moisture to provide for transpiration from the top of the plant. The important role of nutrients in the recovery of a transplanted plant is closely associated with the absorption of moisture. Treatments or conditions that reduce the rate of water loss from the top by transpiration, and enable the root system to absorb water and nutrients, more readily increase the chances of survival of the plant.

Methods of Moving Plants. Three general methods are used in moving plants:

Bare-rooted. One common method of moving horticultural plants is known as bare-rooted transplanting. By this method the root system is removed from the soil in which it has grown, and is replanted in a new location. The root system of a plant moved in this way is seriously damaged by physical injury, and it is subjected to some exposure, both of which are likely to destroy root hairs and growing root tips and to handicap the plant in renewing growth. Nevertheless, this method is used widely for herbaceous plants and for deciduous trees and shrubs.

Shifting. Plants may be moved also by *shifting,* an operation whereby plants are started in pots or similar containers, and from these moved to a larger container or to a permanent location. By this method the soil remains intact, with little or no damage to the root system. This is a means whereby species that do not stand transplanting well are successfully moved.

Balling and Burlapping. Practically the same results as *shifting* may be obtained for larger plants by *balling and burlapping.* In doing this, the plants are dug to include the main roots intact in a ball of earth, which is supported by burlap. This procedure is commonly used in moving evergreen plants, as described later, and also deciduous species during the growing season.

Herbaceous Plants. Many vegetable and flowering plants are transplanted when in a tender, succulent, growing condition. The success with which such plants can be transplanted depends on several factors.

Formation of New Roots. Plants of some species do not stand transplanting well. This is true of corn and many of the peas and beans. It is true also of plants of the cucurbits, such as the watermelon, cantaloupe, and squash. These plants are difficult to transplant because they form new roots slowly and because the roots early develop a suberized layer which makes them ineffective in the direct absorption of water. Root hairs are largely lost in transplanting; and, except under most favorable environmental conditions, the plants can be moved satisfactorily only by shifting. On the contrary, many herbaceous

FIG. 140. Cabbage plants: (1) growing in clay pot, and (2) removed from pot and ready for field planting.

plants can be transplanted readily. This is true of such common vegetables as the tomato, pepper, cabbage, cauliflower, lettuce, onion, and others. It is true also of many flowers as, for example, zinnia, petunia, cornflower, phlox, and verbena. These plants are easy to transplant, apparently because they form new roots quickly and are, hence, soon able to supply the top with moisture. This characteristic is especially noticeable in tomatoes; recently transplanted plants will often form new roots by the second or third day following transplanting. In moving a plant from one location to another, it is desirable not only that the plant live, but that it renew growth as quickly as possible.

Hardening. Strong, stocky plants that have been properly hardened in the seedbed stand transplanting better than soft, succulent plants. Hardening occurs when the growth of plants is retarded. It is accomplished principally by (1) subjecting the plants to relatively lower temperatures, by (2) withholding moisture, and by (3) applying solutions of certain chemicals, such as nitrates and chlorides of potassium, sodium, and calcium. Neither of these treatments should be carried to the extreme, lest the plants be dwarfed severely. The object of hardening is to check the growth of the plant to the extent that it may be able to stand adverse conditions after transplanting to the field, such as higher or lower temperatures, wind, dry soil or air, and hot sunshine. In the process of hardening, the water content of the plants is reduced, and the osmotic concentration increased correspondingly. This condition makes them more retentive of moisture, which is the primary requisite for hardiness to cold, heat, or drought. Hardened plants have a better developed root system in comparison to top growth than is the case with highly succulent plants. Such a root system obviously is able to supply the top more adequately with moisture than is that of a plant not hardened. Hardened plants are more adequately supplied with stored food reserves, which aid the plant in the development of new roots and thereby enable it to become established more readily. Stored food also enables the plant to endure longer before it is weakened by respiration to a point where it can no longer respond. Furthermore, hardened plants do not lose water by transpiration so rapidly as those not hardened. Hardening is desirable even for plants that are to be shifted from pots to the field.

Care in Handling. The care exercised in handling herbaceous plants determines, in a large measure, their response following transplanting. They should be removed from the seedbed with as much of their root system as is practicable and replanted with the least possible delay. They should be protected in the meantime by wet sacks, damp moss, or some other moist insulating material. Often the roots are "puddled," an operation whereby the roots are dipped in a thick mud in order to protect them from excessive drying while they are exposed. In replanting they should be set slightly deeper than they stood in the seedbed, the soil should be pressed firmly about the roots, and water should be added to settle the soil and increase the amount of available moisture.

Weather Conditions. The rate of transpiration is relatively low on cool, moist, cloudy days. The same process normally goes on more slowly late in the afternoon and during the night than during midday. Water requirements are hence less, and the injured root system is able to supply the top more adequately than would be the case if the plant were using more water in transpiration. Thus, plants have a better chance to survive if moved late in the afternoon or on days that are still, cool, cloudy, and humid.

Deciduous Trees and Shrubs. The grape, walnut, peach, fig, and rose are examples of deciduous plants. Bare-rooted transplanting is the method commonly used in transplanting these plants, and the recovery and renewed growth of them is influenced largely by the extent to which transpiration is controlled, by the ability of the plant to develop new functional roots, and by the amount of reserve foods present in the plant.

Transpiration. Most transpiration goes on through the leaves; therefore a logical time to transplant deciduous plants is during their *dormant* period. It is true that the tree is expected to produce new shoots and leaves when it resumes growth the following spring, but in the meantime it will usually have developed new roots sufficient to supply the entire plant adequately with moisture. It is customary to *cut back the top* of trees and shrubs so as to reduce the amount of foliage produced and thus restrict transpiration to an amount likely to be supplied by the root system; or in growth the same results can be obtained by partial or complete *defoliation. Coating the top* of the tree with melted paraffin, paraffin emulsion, or similar preparations reduces evaporation and the consequent weakening of the top.

New Root Formation. Moisture essential for top growth of plants is absorbed largely by *root hairs* or other very minute feeder roots. These are ordinarily destroyed when the tree or shrub is removed from the soil. Furthermore, the tips of small and large roots, the regions from which *feeder roots* arise, are destroyed. Thus, new branch roots must arise before new feeder roots can develop. Ordinarily, new branch roots arise from the pericycle of the portion of the root making primary growth, that is, near the growing tips. These root tips, however, are usually completely destroyed in bare-rooted transplanting, and any new roots that form must then necessarily develop from the cambium of older roots—those that are making secondary growth. These are said to be adventitious roots. Some *kinds of plants* produce such roots readily; others less readily. Differences in formation of adventitious roots possibly account for the ease of transplanting the peach and the difficulty encountered in trans-

Fig. 141. Showing response of pecan roots to chemical treatment prior to transplanting. (1) Toothpicks soaked in indolebutyric acid inserted in roots at points where clusters of new roots occur. (2) No treatment.

planting the pecan. There is some evidence to indicate that adventitious roots normally form more readily on *small roots* than on larger roots of the same plant. Small roots, however, suffer more from drying and other injury. Root pruning during the growing season before a plant is to be moved results in more root branching and a more compact root system. In digging such plants a greater portion of the root system is obtained than is likely when the practice is not followed.

Certain plants produce adventive roots more readily as the

buds begin growth and leaves are formed in the spring. The walnut, pecan, and persimmon are examples of plants of this class. The best time for transplanting them is in very late winter or early spring, when there will be the least delay in initiation of root development.

The formation of adventitious roots may be encouraged or hastened by the use of certain *chemicals*, notably indoleacetic and indolebutyric acid, applied in various ways. Indolebutyric acid has been used successfully in encouraging new root formation in the pecan. Holes are bored transversely into the tap and lateral roots, and toothpicks which have been soaked in a solution of the acid so that they each contain 4 milligrams, are inserted in the holes. Roots form much more readily at points receiving these treatments than at other places.

New root formation takes place most readily in a *well-aerated soil.* If the soil where the tree is planted is kept waterlogged by rainfall or by excessive irrigation, new root formation is discouraged and the plant is likely to suffer.

Reserve Foods. Plants that have made a normally vigorous growth in the nursery stand transplanting better than those that have made restricted growth because of a better supply of reserve-food materials. The reserve food encourages a readier formation of adventitious roots and better top growth, and it supports respiration of the plant more adequately in the meantime.

Two rather distinct practices are followed in the replanting of trees. According to one, the tree is placed in the hole slightly deeper than it stood in the nursery. Loose soil is added and pressed firmly about the roots, which are adjusted from time to time in their natural position as far as possible. Sod, clods, and subsoil encountered in digging the hole should be used last in filling in around the tree and should not be packed in around the roots. It is not advisable to add manure or fertilizer to the soil around the recently transplanted tree.

According to another practice, the soil is shoveled in around the roots of the plant, without any effort to pack it. When the hole is almost filled, water is added to settle the soil, after which the rest of the hole is filled. In either case it is important to handle the tree so that the root system is protected against drying or freezing. Puddling the root system with thick mud, as described for herbaceous plants, is a convenient way to protect it against drying.

Evergreen Trees and Shrubs. Plants that retain their foliage throughout the year are known as *evergreens.* There are two principal kinds. The rhododendron, box, avocado, certain species of ligustrum, and citrus are examples of the so-called *"broad-leaved"* evergreen plants. The pines, cedars, junipers, firs, and arborvitaes are examples of *coniferous* evergreens. In each of these kinds, because of the presence of leaves on a plant, the rate of transpiration is far greater than it is without them, and the moisture required to keep the plant alive is correspondingly greater. Evergreen plants are rarely moved bare-rooted, because in most instances the moisture lost by transpiration from the leaves is greater than can be supplied by the injured root system. Death is inevitable if such a condition exists for very long. Two courses of action may be followed:

FIG. 142. Evergreen shrub, balled and burlapped for transplanting. Burlap is wrapped tightly around top to facilitate handling and to protect the branches.

1. The plant may be *defoliated* in order to lessen the water loss by transpiration and consequently the amount required of the root system. This was formerly an established practice in transplanting citrus, despite the temporary additional dwarfing effect that it had. Although such a practice would be an effective aid in transplanting certain evergreen ornamentals, even temporary defoliation would be objectionable. Such plants are used for their immediate effect in the landscape,

and it is desirable that they retain their original appearance. The use of wax emulsions, previously mentioned, has been found of great value in transplanting such material.

2. The other recourse is to move the plant with a minimum of disturbance to the root system, so that it will continue to function after the plant is moved to its new location. This is done by digging the plants with a portion of the root system undisturbed in a solid ball of earth, which is enclosed tightly in

FIG. 143. Citrus trees and palms in tree shed for transplanting.

burlap. This operation is practiced widely in commercial work and is known as *balling and burlapping* (B&B). The ball of earth seldom contains all the root system of the plant being moved. Some plants naturally have compact root systems and most of the roots can be included in a ball of reasonable size. Others have long, branching roots that are more difficult to include. Obviously, the size of the plant determines how far out its roots extend from the base. In very poor, sandy soils, roots are likely to be longer and less compact than in a fertile soil. Preliminary root pruning is used effectively in encouraging a compact root system. In all cases, however, effort is made to include enough of the roots in the ball to supply the top with moisture during the period when it is becoming reestablished.

Balled stock may be stored for weeks or months before it is replanted. In the meantime, it should be protected against freezing and particularly drying. Balled stock should always be

FIG. 144. Large tree after transplanting. Such trees are pruned moderately and braced securely.

handled in such a way that the ball of earth is not crushed or broken. When plants are stored in this manner for a considerable period, the sacking deteriorates and the balls often crumble, so that a great amount of reworking becomes necessary.

In replanting it is not necessary or desirable to remove the sack. If the soil is packed closely about the ball, the sack will soon decay. The root tips that are not lost or injured continue to grow and give rise to additional feeder roots and to new branch roots that arise in a normal way from the pericycle. The feeder roots in the ball of earth should continue to function in the absorption of moisture as they did before being dug up. Roots, the tips of which are broken, may give rise only to branch roots that are of adventitious origin, and these are not formed so freely or quickly by some species as those that arise in normal succession. All these treatments are closely associated with maintaining for the top a supply of moisture sufficient to prevent permanent wilting, despite a high transpiration rate occasioned by green leaves.

Moving Large Trees. Very large trees are frequently moved by a process comparable to balling and burlapping. Special transplanting equipment is used, and elaborate precautions are taken to move the plant with an appropriate part of its root system intact and undisturbed in a mass of soil around the base. When this is done, it is perfectly feasible to move trees, either deciduous or evergreen, that may be 6 inches or more in diameter.

Treatments that restrict evaporation, such as spraying the top with paraffin emulsion or similar compounds, are considered to be of value in moving large trees. In some cases it is desirable to reduce the top to some extent in order to lessen the amount of water required of the root system.

QUESTIONS

1. Outline briefly the factors that ultimately determine whether a plant lives or dies after it has been transplanted.

2. In what specific way does hardening enable a plant to become reestablished more readily following transplanting?

3. What are the steps in the recovery of a plant that is transplanted bare-rooted? What are the steps for a plant that is shifted or balled-and-burlapped?

4. Why do some deciduous plants resume growth more readily than others after having been transplanted?

5. What various treatments and conditions are favorable to the reestablishment of the root system in the new location?

6. What various treatments and conditions restrict transpiration after transplanting?

7. What is the relationship of reserve stored food to the recovery of a transplanted plant?

SUGGESTED REFERENCES

Cummings, M. B., and R. G. Dunning: "A Study in Recovery of Transplanted Apple Trees," *Vermont Agr. Expt. Sta. Bull.* 432, 1938.

Loomis, W. E.: "Studies in the Transplanting of Vegetable Plants," *Cornell Univ. Agr. Expt. Sta. Memoir* 87, 1925.

Milbrath, J. A., Elmer Hansen, and Henry Hartman: "The Removal of Leaves from Rose Plants at the Time of Digging," *Oregon Agr. Expt. Sta. Bull.* 385, 1940.

Milles, E. J., V. R. Gardner, H. G. Petering, C. L. Comar, and A. L. Neal: Studies on the Development, Preparation, Properties and Application of Wax Emulsions for Coating Nursery Stock and Other Plant Materials, *Mich. Agr. Expt. Sta. Tech. Bull.* 218, 1950.

Romberg, L. D., and C. L. Smith: Effect of Indole-3-Butyric Acid in the Rooting of Transplanted Pecan Trees, *Proc. Am. Soc. Hort. Sci.*, 36:161–169, 1938.

Talbert, T. J.: Transplanting Fruit Trees, *Missouri Agr. Expt. Sta. Bull.* 245, 1927.

Whitten, I. C.: An Investigation in Transplanting, *Missouri Agr. Expt Sta. Research Bull.* 33, 1919.

CHAPTER 16

Pruning

Pruning consists of the removal of certain parts of a plant, to produce some definite modification of the portion that remains. Probably no art of plant craft, except cultivation, is practiced more generally in the management of fruits, vegetables, ornamental plants, and to some extent of forest trees as well.

There are many popular misconceptions about pruning, especially the belief that it is "good for the tree." It should be borne in mind, however, that most types of pruning are contrary to nature, and also that the practice nearly always has a dwarfing effect on the plant. Many experimental studies have shown that the total weight of plant, the diameter of the trunk, and the height and spread of the top are all reduced in proportion to the severity of pruning. This does not mean that pruning is not advisable or necessary, but simply that the reason for pruning must be clearly understood, and the objective in each case must be kept clearly in mind. If this is not done, pruning may be a costly operation in its effect on growth and yield, in addition to the cost of the labor involved.

Under various circumstances roots, stems, leaves, flowers, and fruit are pruned from plants in order to achieve specific objectives.

Root Pruning. The root system is an integral part of the plant, and any treatments, including pruning, that affect the root system should be considered in the light of the influence which they will have on the entire plant.

To Affect Plant Growth. Root pruning by whatever manner it is practiced is a dwarfing process. Cultivation of plants inevitably causes some root pruning, and consequently some

272

dwarfing. The extent of such dwarfing is determined largely by the kind of tillage, the depth and spread of the root system, and the kind of plants subjected to the operation. Likewise, when ditches or excavations are made near plants, these are dwarfed to an extent that is determined by the amount of injury to the root systems. Oftentimes serious damage can be avoided by

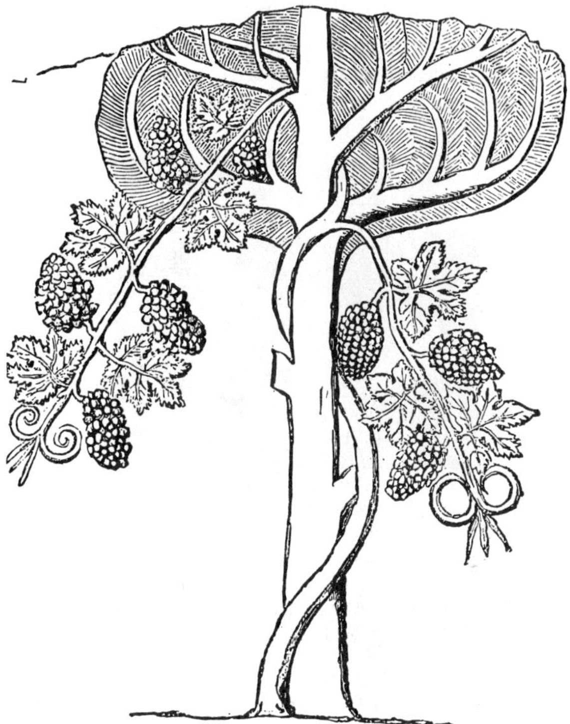

FIG. 145. The ancient practice of training vines on trees, from the Nortn Palace, Koyunjik.

careful planning of such operations, by cutting the minimum number of roots, and by recovering exposed roots without undue delay to prevent injury by possible low temperature and by drying.

Root pruning is practiced purposely on certain kinds of plants to keep them permanently dwarfed. There are several reasons for this practice. Some plants are more attractive and useful as ornamentals if they are kept small in size. The spraying of dwarf trees and the harvesting of fruit from them is easier than the same operations for standard-size trees. Under certain conditions, dwarf plants are proportionately more fruitful than those that grow more vigorously, and root pruning then is one of several ways of causing the dwarfing effect that results in greater fruitfulness.

To Cause Root Branching. Nursery tree rootstocks of pear, apple, and certain other plants are commonly grown by wholesale nurseries for sale to retail nurseries for replanting. Two general kinds of trees are supplied—those with *branched* roots

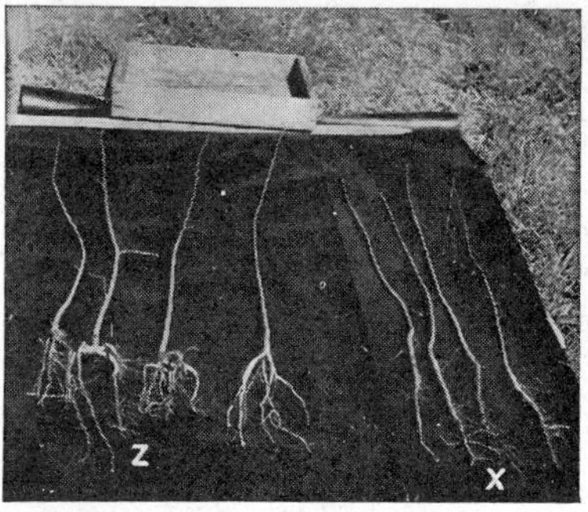

Fig. 146. One-year-old pear seedlings, showing unpruned taproots at X, and branched roots at Z, caused by pruning the tap roots.

and those with *unbranched* taproots. Most buyers prefer the branched roots, because they are easier to replant and they become reestablished more readily. The nurseries that produce the seedlings stimulate the branching by cutting the taproot of the seedling about 3 inches below the ground line, when the plant is from 4 to 5 inches high. The cutting is done by a sharp knife, attached to a tractor and set at the proper angle. Irrigation water is applied promptly to stimulate recovery and renewed growth of the plants. Plants with branched roots produced in this manner sell for a higher price than those with normal taproots. The increased cost results from the expense of cutting the roots and also because the number of marketable rootstocks from a given area is less where they have been dwarfed by root pruning.

To Produce a Compact Root System. Root pruning is frequently performed on plants to encourage the development of the root system in a restricted area. When this has been done, a greater part of the root system may be dug and removed with relative ease, at some future time when the plant is to be transplanted. In actual practice root pruning is done by digging a trench, or by sinking a spade around each plant, to cut the roots at the desired distance; or by using sharp blades mounted on tractors. In any case, the roots respond by branching profusely at the ends where they have been cut and thus produce a more compact root system, which will permit digging and transplanting the plant with better chance of survival.

In Digging for Transplanting. Root pruning is an inevitable procedure in the transplanting of plants bare-rooted or by balling and burlapping. The taproot and the principal lateral roots are usually severed when the plants are dug. As the plant resumes growth, branching occurs where the roots were severed. The normal pattern of root development is thus altered. When the roots resume growth, a plant may have one or more branch roots that grow downward instead of the one central taproot. The tendency, however, is for plants to develop the permanent root system laterally, instead of downward, when the taproot has been severed. Cutting lateral roots likewise causes the development of branch roots in a more concentrated area at the points where the roots were severed.

Top Pruning. Pruning the main body of a tree or shrub is done to influence either the shape or the fruiting response of a plant.

Heading Back. When individual shoots or branches are cut so as to remove the terminal part, the treatment is called heading back. The effect is to stimulate the growth of the remaining lateral buds or branches. Hormones which inhibit the growth of lateral buds are contained in the terminals of shoots; and when these terminals are removed so that the inhibiting influence is no longer present, the lateral buds then grow more freely. In practice, limbs may be headed back drastically or lightly, depending upon the length of the limbs and other factors. Many blind nodes occur immediately below the terminal buds of peach limbs. In pruning peach trees it is customary to head back these limbs to eliminate the area containing most of the blind nodes. The continued practice of heading back will result in a dense, compact plant. Shearing of ornamental plants is essentially a heading-back process and it is practiced widely to produce plants with dense and compact growth which makes certain plants more attractive.

Heading back if judiciously practiced can be used to direct the growth of a tree or plant. If the center of the tree is being closed in, the limbs can be directed outward to provide more light; if lower limbs grow too close to the ground, they can be directed upward by cutting to side limbs that grow upward. Certain chemicals, of which maleic hydrazide is an example, are being used to discourage terminal growth and encourage growth of lateral shoots. The effect is essentially the same as heading back. This treatment has been used successfully on broccoli to encourage the development of small sideheads which are more desirable for the frozen pack than the large central head. It is also an effective treatment to retard tip growth of pyracantha plants, with the resulting stimulation of lateral growth and the development of more compact plants.

Thinning Out. This consists of removing certain branches or large twigs entirely while leaving others nearby undisturbed. Thinning out of certain limbs eliminates competition for those that remain. Growth continues from the terminals so that the limbs get longer each year and ultimately a long rangy type of growth is produced.

It can thus be observed that plants that normally grow thick and bushy, like the apple, will need considerable thinning; while those that grow long and slender, like the peach, will benefit from heading back of the new shoots. In most cases, it will be necessary to use a combination of the two methods, in order to produce the desired effect.

In both heading back and thinning out, the cuts should be made so that they will heal readily. In each case, healing occurs by the growth of tissue from the cambium at the margin of the wound. The healing process, then, is by overwalling.

If cuts are made so as to leave stubs without buds, these do not heal properly but remain weak, and finally die and decay before healing by overwalling can proceed. In heading back, therefore, the main limb should always be cut immediately beyond a node at which a side limb or a lateral bud occurs.

Continued growth of the limb or growth of the bud provides for the quick healing of the cut surface. In thinning out branches, it is proper to make clean smooth cuts, flush with the surface of the limbs from which they arise, thus averting blind stubs.

The Season. Pruning may be done at various seasons on different kinds of plants. If herbaceous plants, like the tomato, are to be pruned it must be done during the growing season. Young fruit trees and ornamental plants are frequently pruned during the growing season to provide the proper form at an early date. The best time for pruning bearing deciduous trees is during the dormant season; for evergreen trees, it is during periods of restricted growth, usually during the fall or winter.

The reason for this practice is that plants normally have a good supply of stored food at the end of the growing season. This stored food enables the tree to resume growth and recover more quickly from the dwarfing effect when dormant pruning is practiced than when pruning is done during the growing season. It also enables wound healing to proceed more promptly, the most favorable time from this standpoint being late winter or early spring. Since a tree uses largely its reserve stored food in starting growth in the springtime, it is left in a weakened condition if pruned during the growing season, with the consequent removal of leaves and the lowering of the food-manufacturing capacity of the tree. Trees that have been weakened by pruning or any other cause are much more susceptible to attack by certain insects, particularly wood borers. The removal of leaves also exposes the limbs and trunk of a tree, and frequently results in the killing of the living tissues on the exposed parts by sunscald.

Treatment of Wounds. The material used for the treatment of wounds must be one that will permit callus growth and wound healing without injury by the wound dressing. Large wounds require a longer time for healing and therefore benefit more from a wound dressing. Fungi which invade open wounds are more destructive in regions of high humidity or heavy rainfall. Thus, wound dressings would be more useful for large wounds and in humid regions. Many commercial preparations are available, and some are entirely effective in protection against decay. One very suitable dressing can be prepared easily by heating together 8 parts, by weight, of resin, and 3 parts, by weight, of raw linseed oil. It is applied like paint, being kept warm to maintain proper consistency.

Fig. 147. Tomato pruned to a single stem.

Shapes and Forms. The grower is concerned with the shape or form of his plants for several reasons. Ornamental plants can be pruned to make them more attractive in appearance. The shape or form of fruit trees, and even some vegetable crops, determines the ease of cultivation, spraying, pruning, and harvesting. There is also a relationship between the shape of a tree and the coloring of the fruit, the susceptibility of the tree to sunscald, and the destructiveness of insect pests and diseases.

Single Stem and Multiple Stem. The most common practice is to prune plants to a single stem at the base. This is true for fruits such as peaches, plums, apples, pears, pecans, and walnuts. Obviously, cultivation of orchard trees is facilitated if the trees are trained to single stems. Some plants, on the contrary, cannot easily be trained to be a single stem. The habits of growth of blackberries and other bramble fruits make it impractical to train to a single stem, and hence several stems are allowed to grow from the base. Fig trees are usually trained initially to a single stem. If perchance they are killed back by cold and later sprout out, they are allowed to develop in bush form with several stems from the ground level. Many ornamental plants produce their best effects when trained to multiple stems. Bush roses are graded according to shape of plant and size of stems. A minimum of three strong canes or stems arising near the ground level is required for a No. 1-grade plant.

Where tomatoes are to be staked, it is necessary to prune the plants. They may be pruned to a single stem, two stems, or even three stems. The number of stems to be left on each plant will determine the spacing to be provided for the plants, and the manner of staking. The tomato plants that are to be pruned can be planted closer together than those that are to be grown without pruning. The closer spacing will provide a larger number of early clusters of fruit from a given area, and hence a greater yield of early tomatoes.

High and Low Heading. Heading refers to the practice of causing, or allowing, several scaffold limbs to develop from the main trunk of the tree. The point at which this occurs is quite high in forest trees where the low branches are continually shaded out. For fruit trees, cultural operations, spraying, and harvesting practices, and the influence of sunlight and ventilation on the tree or fruit determine the height at which they are headed. Peaches and plums must be harvested according to a very definite schedule, and it is easier to harvest the fruit if the greater part of it can be picked from the ground. There is also less likelihood of damage to the trunks of the trees by sunscald if they are headed low. These trees, then, are customarily headed from 18 to 30 inches above ground level. There is less moisture loss from the soil where it is shaded closely by the trees, which is an important factor when irrigation is practiced. Close shading of the ground discourages the growth of competing weeds and grass.

The pecan and walnut are examples of orchard trees that are headed from 4 to 6 feet high. The high heading facilitates ease of cultivation; it provides for better air movement in the grove, and this is important in disease control; it makes it easier to use mechanical equipment in the harvesting of the crop.

In developing a high-headed tree, the lower limbs would not be removed at time of planting but first restrained by heading back, and later progressively removed over a period of years. A tall slender tree with no side branches is subject to sunburn, and the trunk will become inclined permanently by the prevailing wind. Whatever height is originally established for the heading of a tree becomes the permanent height since the trunk does not elongate after secondary growth is initiated.

Type of Framework. The arrangement of the scaffold branches upon the trunk, both horizontally and vertically, will determine whether the framework will be designated as open-center (or vase-shaped), delayed-open-center (or modified-leader), or central-leader. There are certain advantages and dis-advantages of each type.

The *open-center* tree is one where three or possibly four branches are developed with as uniform radial distribution as possible, but in a narrow vertical spacing 18 to 24 inches above

FIG. 148. Hedging machine used to cut back citrus trees to provide more open space between rows. (*Courtesy of Fla. Agr. Exp. Sta.*)

the ground. The tree formed in this way has a low head and is trained to a rather open center, so that it is easily reached for all orchard operations such as spraying, thinning, and harvest-ing. The crotch of such a tree is likely to be weak, and splitting may result in ice storms, windstorms, or even with a heavy load of fruit, unless training has been done with care to pro-duce wide angles. It is, however, the easiest method of training, and certainly the most common, especially with peaches and other stone fruits.

The *delayed-open-center* or *modified-leader* tree is developed with the ideal of a straight trunk 30 to 36 inches in height, upon which four scaffold branches (in case of the peach) are developed at different levels and in different directions radially. It will frequently require two seasons to obtain the branches in

FIG. 149. Hedging machine, shown in Fig. 148, in operation. (*Courtesy of Fla. Agr. Exp. Sta.*)

the desired locations, and there is the objection to the method that is more troublesome to complete than the open-center method. If branches are trained to make a wide angle with the trunk, a strong union is produced so that there is less breakage. This type of training is used quite frequently with apple and pear, and to some extent with peaches and other stone fruits.

In training both open-center and modified-leader trees, where a considerable amount of pruning is involved, it is always recommended that the scaffold branches be selected in early spring, as soon as the new shoots appear on the young tree after it is planted. Other branches should be pruned off as they develop during the season, and the growth of the permanent framework branches favored from the outset. If training is continued in this way, much time will be saved in bringing a tree up to profitable size for bearing. Pruning is traditionally a winter opera-

Fig. 150. Framework of modified leader peach tree, showing proper spacing and strong union of branches with trunk. (*Courtesy of Professor Nino Breviglieri, University of Florence, Florence, Italy.*)

tion; but if it is delayed until the winter following the planting of a young tree, a considerable amount of growth will have to be removed which otherwise would have gone into the permanent framework.

The *central-leader* type is one in which the main trunk is allowed to continue its terminal growth indefinitely, giving rise to side branches in its growth through the years. Trees that are trained by this method have the greatest spread of branches at the base and the least at the top.

Special Forms. Many ornamental plants are commonly trained to special forms and shapes to suit the special preference of the owner. It is common to see hedge plants that have been pruned

Fig. 151. Peach tree trained to a single trunk with three main framework limbs.

so that they are flat-topped, and frequently the sides are sheared vertically.

The *espalier* system is occasionally used in training fruit trees. A novel method used with certain fruits, principally pear and apple, involves training the tree fan-shaped, usually against a wall. A framework is necessary for support at first, but when the branches become hard and woody, they will retain the shape into which they have been trained.

Grapes are trained to special shapes and forms to keep the vines in a shape and to a size that make it easy to maintain them on trellises, which is a necessary procedure in commercial grape growing. Pruned vines produce larger and more compact branches of grapes than unpruned vines. Spraying is a nec-

Fig. 152. Tree in foreground pruned to a high head.

essary practice in grape growing, and vines that are properly pruned can be sprayed more effectively and more economically than unpruned vines.

Grapes may be trained on a *Munson* trellis, whereby each vine has a single stem to a height of 3 feet and two arms trained in opposite directions along the lower wire of a three-wire trellis. Side branches from these two arms are trained outward and upward onto the two wires of the trellis. This system of pruning and training is designed to provide maximum shade for the bunches of grapes, particularly during the ripening period. It also affords some protection from birds, since the bunches hang downward into space, with few vines underneath to provide perches for birds. Grapes are also pruned for training on *Kniffin* trellises. These vines have a main stem to a height of 4 to 4½ feet, with two side branches trained in opposite directions along the lower wire of a two-wire trellis, at a 3-foot

Fig. 153. Pear tree trained in espalier in garden of Governor's Palace at colonial Williamsburg, Va.

level, and two additional side branches trained along the top wire of the trellis. Fruiting branches develop along each of the four main side branches, or "arms," and hang loosely from them. They, in turn, are pruned back or tied up if they become too long.

Pruning of grapes and training to an upright post for support is also practiced. This is known as the *standard* type of trellis, and it is used commonly for European grapes.

Pruning to Eliminate Weak Crotches and Diseased or Dead Wood. Weak crotches are particularly objectionable in the main framework of a tree. V crotches are those that occur as a result of two limbs, of about equal size, growing close together, forming a narrow angle where they join. The angle may be so close that in the growth of the two limbs, islands of bark will be trapped between the two, which will prevent normal union of the tissues of the two adjacent limbs. They are weak in structure because the wood fibers of the two in their growth do not

Fig. 154. Grapes trained on Kniffin trellis.

become properly intermeshed to provide strength. The fibers of the two limbs which provide strength to wood follow a parallel course at the crotch, and they are easily torn apart under the stress of wind, the weight of ice during an ice storm, the weight of a big crop of fruit, or even the normal weight of the limb as it increases in size.

In pruning of young trees, V crotches should be eliminated by cutting off one of the two limbs. Stronger unions can be provided by having branch limbs arise from main limbs at a wide angle, so that the branch limb will be supported at its base by the tissues of the main limb that grow around its base. Such a limb may break under stress, but it will seldom split off.

The removal of dead canes from the *bramble fruits* is a routine part of management. The fruiting canes die after they have produced their crop. The cultivation of the plants and the har-

Fig. 155. Grapes trained on Munson trellis.

vesting of future crops are greatly facilitated if these old canes are pruned out regularly.

A disease of pears, known as *fire blight*, is troublesome wherever pears are grown. No completely satisfactory treatment is known, but the damage can be restricted to some extent by pruning out the diseased limbs and twigs as soon as they are detected. Recent work with antibiotics is promising.

Weak-growing limbs are found on most trees, and it is good orchard-management practice to prune them out. They are especially common on peach and plum where terminal growth becomes restricted and growth is diverted to a vigorous side limb. Likewise, the crowding of limbs in an area and the rubbing or crossing of limbs should be corrected by pruning.

Leaf Pruning. Pruning of stems inevitably results in the removal of leaves also. This is true in the pruning of axillary shoots from tomato plants. The entire shoots may be removed or it may be cut back above the second leaf, in which case the further growth of the shoot will be stopped, but the two remaining leaves will help provide plant food for growth and fruiting of the plant.

Fig. 156. A Y fork in the framework of a tree, one part of which split off during a severe ice storm.

There are a few cases where leaf pruning, as such, is practiced. The principal use is in removal of leaves from plants being trained to special forms to stimulate the formation of side branches at places where they are wanted.

Flower and Fruit Pruning. The pruning of flowers and fruit is common practice with horticultural plants. Pinching off excess flower buds of chrysanthemum will cause the remaining flower or flowers to be larger in size. Flower pruning by hand or chemicals is used to prevent fruitfulness of ornamental plants that produce fruits that are a nuisance. Likewise, the thinning of flowers and fruit of orchard trees to prevent overbearing is receiving the increasing attention of fruitgrowers. Naphthaleneacetic acid is being used extensively in thinning apple and pear flowers. A concentration in common use is made by adding from 4 to 6 ounces of the chemical to 100 gallons of water, the actual strength being determined by the variety upon which it is to be used. It is applied by spraying when the trees are in bloom.

Hand thinning of peaches and plums to prevent overbearing and small fruit is a widespread practice. Peaches are commonly

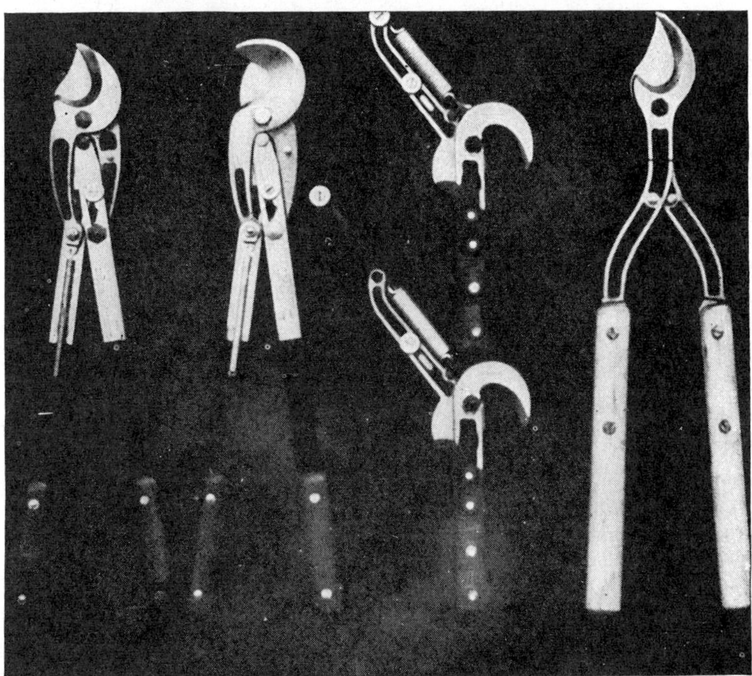

Fig. 157. Different types of heavy-duty pruning shears. (*Courtesy of H. K. Porter, Inc.*)

thinned to a spacing of 6 to 8 inches, during the early development of the fruit.

Grapes are thinned by pruning the younger flower cluster or the small bunches of grapes after blooming. By both of these methods, the entire cluster is removed to improve the nutrition of the bunches of grapes that remain. By these types of pruning, growers attempt to establish a favorable leaf-fruit ratio that will produce regular crops every year, instead of extremely heavy crops one year followed by light crops the following year.

Berry pruning, which consists in cutting off parts of a cluster, either the main stem or the branches, to leave a compact bunch with the desired number of berries, is also practiced. This pruning of the bunches should be done shortly after the berries have set.

Fig. 158. Ornamental plants pruned and trained to special shapes in colonial garden at Williamsburg, Va.

QUESTIONS

1. How can pruning be defined?
2. What are the principal uses or objectives of root pruning?
3. Does pruning make the plant grow larger? Explain.
4. How does heading differ from thinning in practice? In effect on growth?
5. What are the three possible types of framework in trees? Describe briefly.
6. Which type of framework is most commonly used on peach trees, and which on apple trees?
7. When and how are dewberry and blackberry plants pruned?
8. What are the principal methods of pruning and training grapevines?
9. Why do lateral buds on a shoot grow more freely when terminal growth has been discouraged by heading or by chemical treatment?

SUGGESTED REFERENCES

Chandler, W. H., and Ralph D. Cornell: Pruning Ornamental Trees, Shrubs, and Vines, *Calif. Agr. Ext. Serv. Circ.* 183, 1952.

Johnston, Stanley, Ray Hutson, and Donald Cation: Peach Culture in Michigan, *Mich. Agr. Expt. Sta. Circ. Bull.* 177 (1st rev.), 1952.

Luckwill, L. C.: How NAA Thins Apples, *Am. Fruit Grower*, May, 1953.

Magness, J. R., *et al.*: Pruning Hardy Fruit Plants, *U.S. Dept. Agr. Farmers' Bull.* 1870, 1941.

Snyder, John C., and W. A. Luce: Pruning Apple and Pear Trees. *Wash. Agr. Ext. Serv. Bull.* 381, 1949.

Index

Apogamy, 37
Apple propagation (*see* Pome
    fruit propagation)
Apricot propagation, 202
Asexual propagation, 66–68
Avocado propagation, 243–246
    budding, 244–245
    grafting, 244–245
    rootstocks, 244
    top-working, 246

Balling and burlapping, 261, 268
Black end of pears, 218
Blackberry propagation, 235
Blueberry, propagation, 236
Bramble fruits, 234–237
Bud sports, 63–66
Bud union, 159–166
Budding methods, 188–198
    chip, 196–197
    H-bud, 193–195
    patch, 191–193
    ring, 193
    shield or T, 188–191
    skin, 195
Buds, 20–24
    adventitious, 23–24
    axillary, 20
    blind, 21
    classified, 20–23
        function, 22–23
        position, at node, 21–22
            on stem, 20–21

Buds, dormant, 23–24
    fruit, 22–23
    latent, 23–24
    lateral, 20
    leaf, 23
    mixed, 23
    primary, 21–22
    reserve, 21–22
    secondary, 21
    terminal, 20
Budwood, 156–159
    current season, 157–158
    previous season, 158–159
        fresh, 158
        storage, 158–159
Bulbs, 132–139
    aerial, 138
    chemical treatment of, 146
    classification of, 132
    definition of, 134
    layered, 135
    other modified structures,
        139–146
    rest period, 145–146
    scaly, 135
Bush fruits, 236–237

Callus formation, 18–20, 130,
    161
Calyx, 28
Cambium, 18–20
    overwalling, 19–20
    regeneration, 19

Chemical treatments, 84–85
  of bulbs, 146
  of cuttings, 122–124
  of seeds, 80–81, 84–86
  of soils, 59–60
Cherry propagation, 202
Chimera, 167
Citrus propagation, 237–241
  budding, 240
  own-rooted, 241
  rootstocks, 237–239
  training tops, 240–244
Classification of plants, 6–8
Cold frames, 47–53
Compatibility, 162–166
Compost, 56–57
Corms, 139–141
Cranberry propagation, 237
Crown gall, 215
Currant propagation, 237
Cuttage, 110–131
Cuttings, classes of, 110–131
  leaf, 112–115
  root, 110–112
  stem, 115–117
  factors influencing root formation, 119–131
    age of wood, 129
    callus formation, 130
    chemical treatment, 122–124
    etiolation, 130–131
    humidity, 121
    mechanical treatment, 124–126
    media, 119–121
    stored food, 127–129
    temperature, 121
  origin of roots, 117–119

Date propagation, 247
Deciduous plants, 24
Dehorning, 149–150
Dewberry propagation, 235

Dichogamy, 42–43
Diseases, 84–87
  black end of pears, 218
  collar rot, 166
  crown gall, 215–219
  fire blight, 217–219, 287
  root knot, nematode, 60, 200, 226
  on the seed, 85–86
  within the seed, 86–87
  virus, 87
Double-working, 166–167
Dwarf trees, 216, 219–220, 272–273
  apple, 216
  pear, 219–220

Embryo, 35–40
  abortion, 40
  development, 35–36
  polyembryony, 37
Emulsions, wax, 264, 270
Evergreen plants, 24, 267
  broad-leaved, 24
  coniferous, 24

Fertilization, 35–36
Fig propagation, 241–243
  budding, 243
  cuttings, 242–243
  grafting, 243
  seed, 243
Filbert propagation, 232
Floriculture, 4–5
Flower, 26–30
Flower clusters, 26
Flower types, 29–30
Forcing, 46–61, 150–153
  structures for, 46–56
    management of, 56–61
Fruit, 37–39

Fruit, accessory, 38–39
  aggregate, 38
  multiple, 38
  parthenocarpic, 39–40
  seedless, 39–41
  simple, 37–38
  (*See also* specific fruits)
Fruit setting, 41–45

Germination of seed (*see* Seeds)
Gooseberry propagation, 236–237
Graft hybrids, 167
Graft union, 159–166
Graftage, 148–169
  limits of, 164–166
  objects of, 167–168
Grafting methods, 170–186
  approach, 182–185
  bark, 174–176
  bench, 173
  bottle, 177
  bridge, 180–182
  cleft, 170–172
  cutting, 177
  inlay, 176–177
  nurse-root, 178
  root, 178
  veneer, 177
  whip, 172–174
Grafting wax, 186
Grafts, after care of, 178–180
Graftwood, 155–156
Grape propagation, 220–226
  budding, 225–226
  cuttings, 221–222
  grafting, 223–225
  layers, 222–223
  rootstock, 226
  seed, 221
Greenhouses, 53–54
Guava, 246

Hardening, 263
Herbaceous plants, transplanting, 261–264
History, 2
Horticultural industries, 2, 5–6
Hotbeds, 47–53

Incompatibility, 41–42
Inflorescences, 26
Internode, 16

Jujube propagation, 246

Layerage, 100–109
  air, 106–108
  compound, 102–103
  continuous or trench, 103–105
  mound or stool, 105–106
  as preliminary treatment, 108–109
  simple, 101
  tip, 101–102
  uses, 100–101
Leaves, 24
Life cycle, 7–8
Lychee propagation, 246

Materials, waxed tying, 190
  wound covering, 179–180
Matrix, 154
Multiple seedlings, 36–37

Nectarine propagation, 201
Nematodes, 60, 200–201, 226
Nodes, 16–18

Olericulture, 4
Olive propagation, 246

Ornamental horticulture, 3–4
Overwalling, 19–20

Papaya propagation, 247
Parthenocarpy, 39–40
Peach propagation, 199–209
  budding, 203–205
  rootstock, 199–200
  top-working, 207–209
Pear propagation (*see* Pome fruit
    propagation)
Pecan propagation, 228–232
  budding, 229
  grafting, 230
  nursery trees, 228–229
  rootstocks, 228, 231–232
  top-working, 229–231
Pedicel, 26
Peduncle, 26
Perianth, 28
Persimmon propagation, 233–
    234
  budding, 233–234
  grafting, 233–234
  rootstocks, 233
Phylloxera, 223–226
Pistil, 27–28
Plant dissemination, 8–10
Plant propagation, 5, 62–70
  (*See also* specific plants)
Plant protectors, 46–47
Plant structure, 12–25
  buds, 20–24
  leaves, 24
  roots, 12–14
  stems, 15–20
Plants, classes of, 7–8
  annual, 7
  biennial, 7
  deciduous, 24
  dicotyledonous, 17
  dioecious, 29–30, 43
  dissemination, 8–10

Plants, classes of, evergreen, 24
    broad-leaved, 24
    coniferous, 24
    horticultural, 3–6
    introduction of, 9
    life cycle of, 7–8
    monecious, 29
    monocarpic, 7
    monocotyledonous, 15
    perennial, 8
  methods of moving, 261
    balling and burlapping, 261
    bare-rooted, 261
    large trees, 270
    shifting, 261
  propagation of important,
      199–256
    (*See also* Pome fruit propa-
      gation; Stone fruit propa-
      gation; specific plants)
Plum propagation, 201–202
Pollination, 30–35
  agents of, 32–35
  cross-, 30–32
  self-, 30–32
Polyembryony, 37
Pome fruit propagation, 209–220
  apple, 209–216
    budding, 212–213
    clonal rootstocks, 212
    double-working, 215
    dwarf trees, 216
    grafting, 213–215
    intermediate rootstocks, 211
    own-rooted, 215
    seedling rootstocks, 210–211
    top-working, 215–216
  pear, 216–220
    budding, 218–219
    double-working, 219
    grafting, 218
    hardwood cutting, 220
    quince rootstocks, 219–220
    seedling rootstocks, 216–218

Pome fruit propagation, pear, top-working, 219
Pomegranate, 246
Pomology, 3
Propagation, 1, 56–61, 66–70
  beds, 56, 61
  of important plants, 199–256
  methods of, 68–70
  types of, 66–68
    (*See also* Pome fruit propagation; Stone fruit propagation; specific plants)
Pruning, 272–290
  flower, 288–290
  forms, 278–285
  fruit, 288–290
  leaf, 288
  root, 272–275
  season for, 277
  shapes, 278–285
  top, 275–277
  wounds from, treatment of, 277–278
Puddling, 263

Quince propagation, 219–220

Raspberry propagation, 235–236
Receptacle, 26, 38–39
Regeneration, 19
Rest period, of bulbs, 145–146
  of seeds, 83–84
Rhizome, 141–143
Root hairs, 14
Root knot, nematode, 60, 200, 226
Roots, 12–14, 118–119
  fleshy, 144–145
  formation in transplanting, 261–264, 268
    chemical stimulation of, 266
  origin in cuttings, 117–119

Roots, types of, 12–14
  adventitious, 13–14
  branch, 13
  lateral, 12–13
  primary growth of, 12
  primordia, 118–119
  regular, 13
  secondary growth of, 14
  tap, 12
Rose propagation, 247–256
  budding, 251–254
    cane, 253
    dormant, 252
    June, 252
  cuttings, 248–250
    hardwood, 248–249
    softwood, 249–250
  grafting, 251
  layers and suckers, 254
  own-rooted, 254
  seeds, 247–248
  stocks, 250
  tree or standard, 253–254

Scions, 154–155
  types of, 155
Seedage, 92–99
  field, 92–95
  methods for fruits, 98–99
  in special beds, 95–98
Seedling, 36–37
Seeds, 36–37, 71–89
  afterripening of, 83–84
  chemical treatments of, 80–81
    to control disease, 84–87
  classes, 88–89
  coverings, 79–83
  germination of, 71–90
    delayed, 79–84
    methods of hastening, 80
    processes of, 71–72
    requirements for, 73–75
  growing of, 87–88

Seeds, parts of, 36
   rest period, 83–84
   testing, 89–90
Shifting, 261
Slatted frames, 55–56
Soil pasteurization, 58
Soil preparation, 56–57
Soil sterilization, 57
Soils, chemical treatments of, 59–60
Stamens, 27–28
Stems, 15–20
Stock and scion effects, 162–163
Stocks, 153–154
Stone fruit propagation, 199–209
   budding, 203–205
   budwood, 205–207
   cuttings, 203
   handling of seed, 202–203
   rootstocks, 199–202
     apricot, 202
     cherry, 202
     nectarine, 201
     peach, 199–200
     plum, 201
   top-working, 207–209
Stored food, 36, 71, 101, 127–129, 155, 161, 261, 263, 266, 277
Stratification, 82
Strawberry propagation, 234
Suckers, 23, 110

Testa, 36
Top-working, 148–149
   apples, 215–216
   avocado, 246
   peach, 207–209
   pear, 219
   pecan, 229–231
Transpiration, reducing, 264
Transplanting, 260–270
   deciduous trees and shrubs, 264–267
   evergreen trees and shrubs, 267–270
   herbaceous plants, 261–264
   methods, 267
   moving large trees, 270
Tree-nut crops, 227
Tubers, stem, 143–144
Tung nut propagation, 232–233

Varieties, origin of, 62–66
Vegetable gardening, 4

Walnut propagation, 227–228
Water sprouts, 23
Wax emulsions, 264, 270
Waxes, grafting, 186
Woolly aphis, 219